KV-433-440

Glands—The Mirror of Self

Glands—The Mirror of Self

By

Onslow H. Wilson, Ph.D., F.R.C.

Rosicrucian Library
Volume XVIII

SUPREME GRAND LODGE OF AMORC, INC.
Printing and Publishing Department
San Jose, California

FIRST EDITION, 1983

Library of Congress Catalog Card No.: 83-80358
ISBN 0-912057-35-1

Second Printing, 1984
Third Printing, 1989

The Rosicrucian Library

Volume

I	Rosicrucian Questions and Answers With Complete History of the Order
II	Rosicrucian Principles for the Home and Business
III	The Mystical Life of Jesus
IV	The Secret Doctrines of Jesus
V	Unto Thee I Grant
VI	A Thousand Years of Yesterdays
VII	Self Mastery and Fate With the Cycles of Life
VIII	Rosicrucian Manual
IX	Mystics at Prayer
X	Behold the Sign
XI	Mansions of the Soul
XII	Lemuria—The Lost Continent of the Pacific
XIII	The Technique of the Master
XIV	The Symbolic Prophecy of the Great Pyramid
XVI	The Technique of the Disciple
XVII	Mental Poisoning
XVIII	Glands—The Mirror of Self
XXII	The Sanctuary of Self
XXIII	Sepher Yezirah
XXV	Son of the Sun
XXVI	The Conscious Interlude
XXVII	Essays of a Modern Mystic
XXVIII	Cosmic Mission Fulfilled
XXIX	Whisperings of Self
XXX	Herbalism Through the Ages
XXXIII	The Eternal Fruits of Knowledge
XXXIV	Cares That Infest
XXXV	Mental Alchemy
XXXVI	Messages From the Celestial Sanctum
XXXVII	In Search of Reality
XXXVIII	Through the Mind's Eye
XXXIX	Mysticism—The Ultimate Experience
XL	The Conscience of Science and Other Essays
XLI	The Universe of Numbers
XLII	Great Women Initiates
XLIII	Increase Your Power of Creative Thinking
XLIV	Immortalized Words of the Past
XLV	A Secret Meeting in Rome
XLVI	The Mystic Path
XLVII	Compass of the Wise

(Other volumes will be added from time to time.
Write for complete catalog.)

Contents

∇

Illustrations

Preface

There is considerable emphasis today upon the mechanical-electrical nature of man's being. The science of cybernetics is devoted to a comparative study and theory of the mechanical-electrical functions of the brain and nervous systems. Simply, man has developed mechanical and electrical systems similar to those of the human organism, but far exceeding in many regards, their capabilities. Robots are increasingly being developed which perform certain actions with far more efficiency than humans can.

However, man has not yet fully and empirically explained, for example, the nature of *consciousness.* The area of special response of consciousness has been demonstrated, and of course it is essential as an attribute of life. But just what is *consciousness*? Is it a thing, a substance, a function? And further, how did it originate in the simplest life forms? These are still highly speculative subjects.

All organs and systems of the living organism appear to have been conditioned for a special duty or function to perform. In terms of the Computer Age, they would seem to have been *programmed*—by *what*, by *whom*? The D.N.A., located in the cell nuclei, are said to be the molecular basis of heredity. Though they may be altered, there is the question, How did they acquire their original efficacy? Why, in fact, has even the simplest living cell all of the principal functional factors of Life Force?

Dr. Onslow Wilson, in his work, *Glands—The Mirror of Self*, expounds upon the innate directional functions which the glands have over the other organs and systems of the human. It would appear that they are super (psychic) intelligence-control centers. Yet again, though they may seem to mechanically conform to their nature, the mystery of *why* they do so still persists. In other words, why are they so motivated? The subject is explored by Dr. Wilson. As to *how* the glands perform is also simply and effectively explained in this book.

—Ralph M. Lewis

An Introduction to the Study of Glands

By Dr. H. Spencer Lewis

The Rosicrucians have maintained for centuries that the glands act like *guardians* of the lives of human beings.

When we stop to realize that man in his earthly existence is functioning as a dual being, and that there is a spiritual self within a physical body, and that the spiritual self is there for the purpose of giving man intellectually a sense not only of divine wisdom and divine mastership over earthly conditions, but to guard and control the perfect operation of the physical body, we must realize that there must be also some means of exchange or communication between the spiritual self and the physical self. In other words, there must be some places or points within the human body where the spiritual power, self, and intelligence can transmute its power, authority, and control into the grosser elements of nerve energy, blood, vitality, and human mechanism so that the higher, finer, almost intangible and imperceptible forces of the divine self may be brought down to a rate of vibrations and a form of power crude enough, or material enough, to function through the flesh and bones and other material, chemical elements that constitute the body of man.

The glands have been found to be these intercommunicating instruments, these transformers, or transmuters between the spiritual, divine, Cosmic self and the grosser, earthly and physical self. They bring about within man a *divine alchemy.* For many centuries the most eminent mystical scientists, who made a very serious study of the rhythmic, synchronous functionings of both the divine and physical self in man's body, believed that the pineal and pituitary bodies, now known to be glands, were the only actual physical, material organs for such transmutation of a higher force and energy into a more material force. On the other hand, there were those who believed that the solar plexus was a gland of great importance, having the function of interpreting and transmuting the higher, inspirational, Cosmic, or spiritual emotions within man into the grosser, material, and emotional reactions. For a century or more the solar plexus was somewhat worshipped and adored as the seat and soul of all of man's higher activities. But when it was discovered that the spiritual element within man is to be found in every living cell of every part of bone, tissue, and blood, and that the soul and emotional nature of man are not located in one organ or one part of the body, it became necessary to study man's physical anatomy more carefully. Then the many other glands were discovered and given proper attention.

Speaking of the emotional centers of man's body again, we have found, as have scientists and medical men, that the spleen is just as reactive and just as demonstrative of the emotional functions of man's mental, psychic, spiritual, and physical existence as is the solar plexus. This, too, was discovered many centuries ago, and for that reason many popular phrases were invented by the more or less ignorant laymen whereby they expressed the idea that one who was despondent or unhappy or cranky was manifesting a bad spleen. But it is also true that no part of man's spiritual and physical composition can be out of order or out of harmony with the Cosmic rhythm or with the Cosmic flow of vibrations without man's emotions reacting and manifesting the inharmonious attunement.

From many mystical and spiritual points of view the pituitary and pineal bodies or glands may be quite important in certain so-called "psychic" reactions. No one knows better than do the Rosicrucians that these two glands or bodies should be given careful thought in connection with many forms of development of the latent spiritual, or Cosmic, abilities of the human being. But then again there is the thyroid gland which, while it does have a considerable importance in connection with the development and growth of the physical human body, and from the physical, medical point of view may be closely related with certain forms of malignant or toxic conditions that are subnormal or abnormal, on the other hand, is important in certain forms and degrees of psychic or spiritual development.

It is not necessary for every individual to become mystically inclined or to be given to the study and reading of mystical, spiritual, or religious subjects in order to be benefited by a very careful study of the glands within the human body. Our human countenances, our human attractiveness, and most essentially that intangible something called *human personality* or *personal magnetism*, are the result of the normal and proper functioning of the glands. And that which attracts one person to another is something more than the mere definiteness of the handclasp or the deliberateness of the smile, or the wiles of the pleasant words that are spoken.

By knowing our glands and how they function, and by knowing how to live properly, which includes eating, drinking, and breathing properly as well as thinking properly, we can permit these glands to do their very best, and give us every advantage of their divine functioning.

1 The Gonads

There is an old aphorism which states: "One never misses the water until the well runs dry." Indeed, it is only after the well has run dry that we come to appreciate fully the importance of water in our daily lives. In an analogous manner it is only through the malfunction or exhaustion of a gland that we generally come to a fuller appreciation of glandular contributions in the economy of our physical and psychic lives. Accordingly, much of what is known about glandular function has been derived from the study of individuals, animal and human, that have either lost the function of a gland, or are inflicted with some impairment in glandular function. But what are glands?

Two Types of Glands

Glands are organs, consisting of specialized tissues, which produce and/or secrete materials essential to body harmony. Essentially there are two types of glands. One type of gland has ducts through which the special secretions pass on their way to their final destination. Such duct glands are known as *exocrine* glands. Salivary glands, tear glands, and the liver are examples of duct or exocrine glands. The secretions of exocrine glands empty directly on to those body surfaces at which their influence is to be exerted.

The other type of gland is ductless and is referred to as *endocrine.* The word endocrine is derived from two Greek words: *endo* meaning "within" and *krino* meaning "I separate." It appears, therefore, that the word "endocrine" alludes to the inner or secret functions of these highly specialized ductless glands. Unlike the exocrine or duct glands, the secretions of the endocrine glands ooze into the bloodstream from the cells comprising the glands. The blood then transports the glandular elements to the various parts of the body.

The known endocrine glands are: the gonads, the pancreas, the adrenals, the thymus, the thyroid, the parathyroids, the pituitary, and the pineal. Each of these endocrine glands plays a specific role in the economy of our physical life and shares a correspondence with analogous functions in our psychic life. As a consequence, a study of the glands can lead to a fuller appreciation of our psychic self.

The ductless or endocrine glands may be divided into two groups according to whether or not they are essential for the expression of life in the body. Thus one group consists of those glands without which physical life is impossible. To this group we may assign the term "vital." The second group of glands consists of those without which the character of the physical body is so altered as to leave individuals open to types of social experiences which otherwise may not come their way. In some such cases the physical alterations are not exaggerated and, as a consequence, the individual may live a normal, well-balanced life both physically and psychically. In other cases the physical alterations are so exaggerated as to leave little doubt regarding the normalcy both of the physical and psychic functions of life.

To the first or "vital" group of ductless glands belong the adrenals, the pancreas and the parathyroids. Life in the human body is not possible without the secretions of these glands. To the second group of glands belong the gonads, the thymus, the pituitary, the thyroid, and the pineal. Although the secretions from this second group of endocrine glands are not vital to the expression of life in the physical body, they are important to the quality of life as it manifests in the physical body. What then are the glandular secretions, and how do they contribute to the economy of life? Let us begin with the gonads.

The secretions of the endocrine glands are known as hormones. The hormones of the gonads belong to a class of molecules called steroids. The hormones produced by the female gonads or ovaries are known as estrogens, while those produced by the male gonads or testes are known as androgens.

The word gonad is derived from the Greek *gonos*, meaning "seed." An equivalent to gonos is the Greek, *gignomai*, meaning "to become." The gonads may therefore be regarded as organs having to do with the concept of becoming or generation. Indeed, in addition to being endocrine or ductless glands, the gonads are also exocrine or duct glands. As exocrine glands the gonads are not vital to the survival of the individual. That this is so is clearly demonstrated by the fact that there are, and have been, innumerable individuals who have lived and continue to live ordinary, healthy lives without gonadal function. Nevertheless, the gonads do function in supplying the "seeds of becoming" for the species. From this point of view the exocrine functions of the gonads are vital to the survival and progression of life within the species.

Menopause

Perhaps the most common occurrence of loss of gonadal function during adult life comes in the wake of the menopause in both male and female. Although information on the physiological effects of the male menopause is scanty at best, there is evidence which indicates that during the middle years the male tends to be less aggressive in his behavior. In the male the decrease in gonadal function during menopause is generally so gradual that the other endocrine or ductless glands are able to adjust quite readily. As a consequence, from a physiological and behavioral point of view, the male menopause tends to be quite uneventful. On the other hand, during female menopause, the relatively rapid withdrawal of female hormones from the physiological scene often requires an equally rapid readjustment on the part of the other glands. Until a new equilibrium is established, the female menopause is often accompanied by a pronounced sense of uneasiness which, for some, is a source of great anxiety. Menopause in both male and female is accompanied by changes in the metabolic rate and an increased tendency toward obesity. However, there is usually no dramatic change in personality or behavior following menopause.

Puberty

Loss of the gonads before the age of puberty naturally precludes any contribution to the progression of life within the species. In addition, absence of the endocrine functions of the gonads prior to puberty has important consequences for the type of body in which the individual will experience the flow of life. Accordingly, in the male the voice does not deepen, facial and other body hair is sparse, the body is less muscular than normal, the rate of metabolism is lower and there is a tendency to be fat. In the female, loss of gonadal function prior to puberty leads to the development of a more muscular body, deeper-pitched voice, more pronounced body hair, narrow pelvis, little or no development of the mammary glands of the breasts and less of a tendency to put on fat in the region of the hips and upper thighs. In both male and female, the long bones of the body tend to be longer than normal because the growing points of these bones fuse much later than would otherwise be the case.

Mentally and emotionally, individuals who have lost gonadal function prior to puberty tend to differ from the norm. Thus, it appears that in males, withdrawal of the male hormones prior to puberty leads to a more docile, easygoing type of male than normal. Conversely, in the females, withdrawal of female hormones prior to puberty seems to lead to a more aggressive type of personality. However, it is not yet clear as to whether these behavioral differences arise as a consequence of physiological alterations in the brain and nervous system, or whether they arise as a consequence of conscious or unconscious psychological adjustments on the part of the individual to social attitudes consequent to their somewhat abnormal physique.

Genes, Sex Hormones, and Foundations of Behavior

Questions relating to the influence of glandular secretions on behavior and personality are of paramount importance to the student of mysticism because answers to such questions touch directly on the concept of Free Will. Since the functions of the Central Nervous System are the means through which our behaviors and personalities are ultimately reflected and expressed, it behooves us to examine what is known concerning the influence of gonadal hormones on the structure and function of the body in general, and on the Central Nervous System in particular.

Animal behavior is predicated upon three levels of organization. The first level of organization is the genetic. At the genetic level of organization the individual is bound to the species by exhibiting certain fundamental, instinctive, unconscious behavioral characteristics common to all members of the species. The second level of organization is the physiological. The physiological level of organization is an extension of the genetic but is also the medium through which the individual becomes conscious of itself. The third level of organization is the psychological which develops largely as a consequence of the genetic and physiological. The psychological level of organization is the means by which the individual learns to develop and experiment with strategies conducive to the survival of its individuality and, as a consequence, that of the species.

In most animal species individual behavior is largely determined by the genetic and physiological levels of organization. In most species the psychological level of organization is generally of such relatively low development as to have little influence on their physical and psychic natures. In man, however, the situation is quite different. In man, the third, or psychological, level of organization is so well-developed that we refer to it as Free Will. Because of Free Will we are able to redirect the vital flow of life in our physical and psychic beings. Nevertheless, this freedom of will can be exercised only if certain structural requirements of the body are met.

The genetic makeup of the individual body has profound consequences for its structure and consequently its behavior. For example, the foundation for male-female (gender) differences in animal behaviors is laid at the time of fertilization. Fertilization is the process during which a male sex cell unites with a female sex cell to form a zygote or fertilized egg. Every female sex cell, called an ovum or egg, carries the same type of sex determining chromosome, the X-chromosome. On the other hand, male sex cells, called sperm cells, are of two types depending on whether they carry an X-chromosome or a Y-chromosome. As a consequence, when egg and sperm unite, one of two types of zygotes will result. Thus, if the uniting sperm cell carries an X-chromosome, the resulting zygote will carry two X-chromosomes and following implantation in the uterus, will develop into a female embryo. On the other hand, if the sperm cell which unites with the egg carries a Y-chromosome, then the resulting zygote will carry one X-chromosome and one Y-chromosome and will develop into a male embryo. See Figure 1-1. We see, therefore, that the genetic foundation for individual behav-

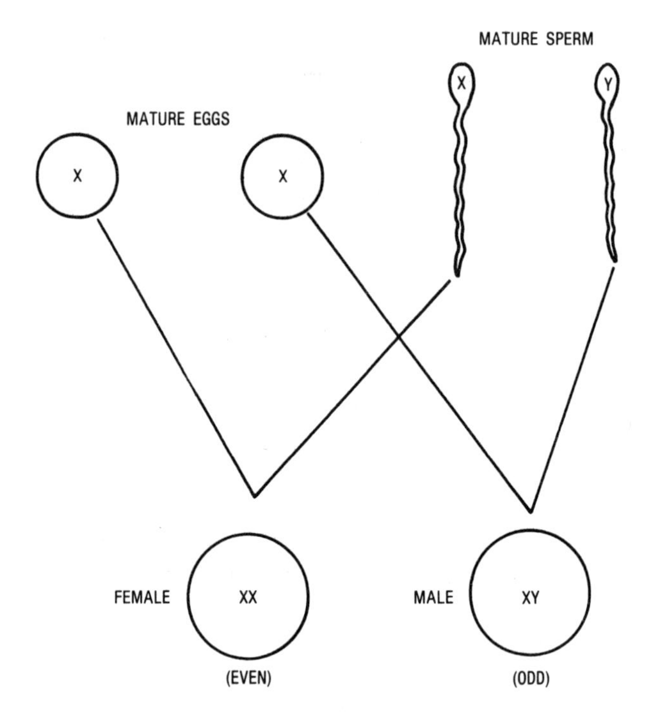

Figure 1-1. Schematic representation of the fertilization of mature eggs by mature sperm, giving rise to male (XY-chromosome carrying) and female (XX-chromosome carrying) zygotes. When zygote is securely implanted in the uterus it develops into a male or female fetus.

ior is determined at the time of fertilization. As a consequence, the framework in which the physiological level of organization is to unfold is established at conception.

In the human, an important aspect of the second or physiological level of organization is established at about the sixth week after fertilization. If the developing fetus is to become a male, then between the sixth and eighth week of gestation, the activities of the male-determining genes located on the Y-chromosome begin to induce the primitive gonadal tissue to differentiate along male lines. The primitive gonadal tissue, known as the genital ridge, is neither male nor female. It consists of an inner portion, or medulla, and an outer portion, or cortex. The inner portion has the potential for development into the male gonad, while the outer portion represents the female gonad in potential.

Between the sixth and eighth week after fertilization the activities of the male-determining genes of the Y-chromosome induce the inner portion of the genital ridge to develop into the male gonads or testes. On the other hand, in the female fetus, the absence of male-determining genes facilitates the development of the outer portion of the genital ridge. As a consequence, in the female fetus, the cortex of the genital ridge develops into the female gonads or ovaries. See Figure 1-2.

Following differentiation both male and female gonads begin to produce their characteristic hormones. Thus, the testes produce the male sex hormones known as androgens, while the ovaries produce estrogens. With the advent of gonadal function all tissues of the developing fetus are exposed to the hormones secreted by these glands. As a consequence, the secretion of sex hormones by the gonads of the fetus establishes the physiological level of organization as far as gender-related behaviors are concerned.

Sex Hormones and the Nervous System

Circulating male sex hormones, particularly testosterone, seem to play a crucial role in determining whether the developing brain and nervous system will differentiate along male or female lines. As a consequence, male sex hormones appear to have a profound indirect influence on behavior and hence on the third or psychological level of organization. For example, in many species there are marked male-female differences in Central Nervous System control of glandular function and, consequently, behavior. Such differences in Central Nervous System function are due, in part, to the influence of gonadal secretions. Thus, experiments with labora-

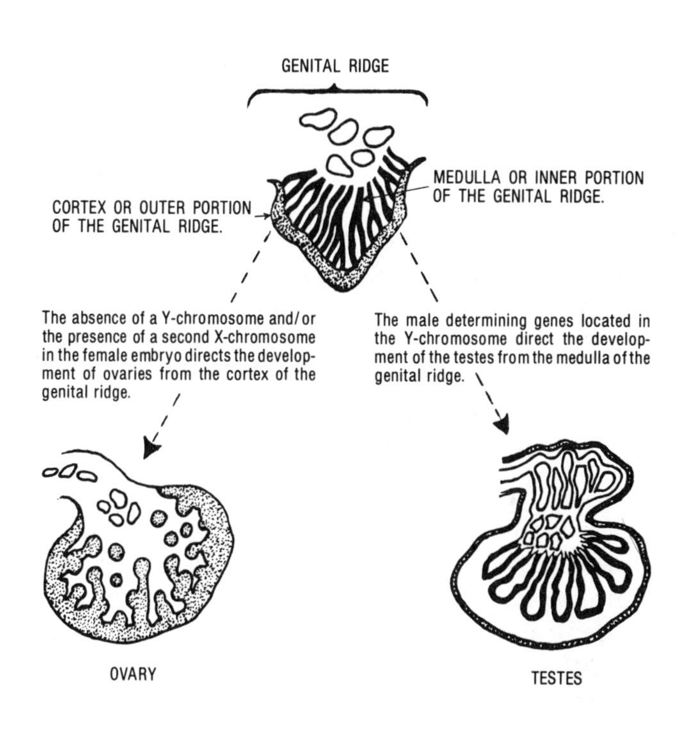

Figure 1-2. Differentiation of androgynous genital ridge into female (no Y-chromosome) and male (Y-chromosome) gonads.

tory rats have shown that the appearance of masculine patterns of behavior and of anterior-pituitary hormone secretion during adult life are dependent upon factors released by the male gonads (testes) during the period shortly after birth. This evidence, coupled with the knowledge that the functions of the pituitary gland are regulated by centers in the brain (the hypothalamus) led scientists to the realization that the testes must, in some way, influence the development of centers located within the brain.

In the absence of a Y-chromosome, the fetus develops along typically female lines. This fact, coupled with the realization that male sex hormones influence the development of certain brain-centers, has given rise to a general model for Central Nervous System (CNS) differentiation. According to this general model, the inherent, inner pattern of CNS development must be along female lines. In other words, in the absence of male sex hormones, the Central Nervous System of mammals follows an inner impulse and develops along typically female lines. As a consequence, functions and behaviors that are dependent upon the structure of the CNS for expression would be characteristically feminine. According to this view, male patterns of behavior and CNS structure must be the result of a departure from the intrinsic female patterns, and must, therefore, be brought about by the action of male sex hormones produced by the fetal testes.

Although this simple mechanism cannot be the only determining factor in male-female differences in Central Nervous System structure and function, there is considerable evidence that, in many cases, early exposure of the nervous system to sex hormones is a contributing factor. Such influences of the sex hormones on the organization and structure of the Central Nervous System set the stage for the psychological level of organization. In this regard it is interesting to note that male sex hormones are not the only sex hormones that induce the types of structural and functional changes in the Central Nervous System that lead to the appearance of masculine patterns of behavior. Female sex hormones are known to exert the same influences. Why then the differences between male and female?

The effects of male sex hormones appear to be exerted directly on the Central Nervous System. Thus, in the rhesus monkey, an animal whose physiology is remarkably similar to that of the human, direct exposure of the brains of female fetuses to androgens during a critical sensitive period produces masculinization of the Central Nervous System. This masculinization leads to the subsequent appearance of masculine patterns of behavior during the adult life of these females. Paradoxically, similar direct exposure of female brains to female sex hormones also results in the masculinization of the female brain and behaviors. The mystery deepens when it is recognized that the brains of mammalian fetuses, including those of humans, are exquisitely sensitive to estrogens. Interestingly, brain tissue is very active in converting male sex hormones to female ones. It thus appears that masculinization is due directly to the action of female sex hormones on the Central Nervous System, and only indirectly to male sex hormones. Accordingly, male sex hormones must first be converted to the female variety before their influence can be made manifest. How then are the brains of mammalian female fetuses protected from the masculinizing influence of their own female sex hormones and those of their mothers?

In many mammalian species the brain of the developing fetus is functionally protected from the masculinizing effects of female sex hormones. This protection is afforded by the presence of a protein in the blood which binds female sex hormones. This female sex-hormone-binding protein is known as alpha-fetoprotein. A similar male sex-hormone-binding protein has not been found. As a consequence, it is believed that the developing brain of the male fetus is not protected from the masculinizing effects of circulating sex hormones.

Although alpha-fetoprotein is not found in human blood, recent evidence indicates that a similar binding protein does exist in the blood of the human female. The circulating levels of this binding protein are regulated by the activities of the thyroid gland. As yet, such a sex-hormone-binding protein has not been found in the blood of the human male. It appears, therefore, that masculinization of the human brain may also be indirectly accomplished through the agency of male sex hormones. Upon reaching the brain, these male sex hormones are converted to female sex hormones which then masculinize the brain during its critical sensitive period.

The Nervous System and Behavior

The question now arises: Does masculinization of the human brain impose a set of unalterable behavioral patterns which must be slavishly adhered to during adult life as is the case with other species? Or does brain masculinization only predispose the individual to certain patterns against which Free Will may prevail if one so chooses? Let us briefly examine the scientific evidence in this regard.

Because the adrenal glands also produce relatively small amounts of the sex hormones, hyperactivity of these glands often results in exposure of the developing female brain to abnormally high levels of male sex hormones. Accordingly, the behavior of girls who were exposed prenatally to high levels of male sex hormones due to such adrenal hyperactivity differs significantly from that of girls who were not. Typically, such masculinized girls demonstrate aspects of behavior interpreted by researchers as masculine. These aspects include: (1) a combination of intense active outdoor play, increased association with male peers, long-term identification as "tomboy" by self and others; (2) a decrease in play with dolls, etc.; (3) an increased participation in body contact sports but no remarkable increase in aggressive behavior.

Interestingly, the behavior of boys, who in fetal life were exposed to abnormally high levels of male sex hormones due to adrenal hyperactivity, differs from that of their male siblings only in that they exhibit increased levels of energy expenditure while engaged in activities such as play and sports. In such cases it seems that prenatal exposure to high levels of male sex hormone results in an inclination toward intense active outdoor play and possibly some aspects of aggressive behavior.

Other evidence with regard to the anatomical and behavioral influences of prenatal exposure to abnormally high levels of sex hormones relates to the use of hormonal interventions during pregnancy. Since the early 1950's millions of pregnant women have been treated with artificial female sex hormones, progestogens and estrogens, in order to minimize the risks of threatened abortions. It has now become clear that these orally administered artificial hormones have important consequences for the developing fetus. For example, not only is behavior masculinized, but it has been shown that these hormones also produce masculinization of the genitalia in as many as 18% of female offspring.

From a structural point of view, it is well established that exposure of brain tissue to male or female sex hormones accelerates and enhances the outgrowth of nerve axons (neurites) in certain regions of the brain including the pre-optic area and portions of the hypothalamus. Here too, female sex hormones appear to be of primary importance, and it is assumed that male sex hormones are first converted to the female variety before their effects can manifest. This influence of sex hormones on neurite growth assumes particular importance in view of the emerging evidence that early exposure of the developing fetus to gonadal hormones actually induces structural alterations within the Central Nervous System.

Learning and Behavior

Certainly the evidence concerning the influence of early exposure of the Central Nervous System to sex hormones and the consequences of such early exposure for the psychological temperament of the individual is compelling. Nevertheless, in observing overall behavior, the importance of learning cannot be overlooked. Thus, in many mammals, early social experience has profound influences on subsequent behavior. In humans the role of this social learning in the development of behavior is much greater than in other mammals. For example, in humans the development of a sense of identity as male or female depends largely on a process of learning. Studies with children who were born with the anatomy of one sex but brought up as a member of the other sex suggest that male-female identification depends on the sex role *learnt* in childhood.

To date the evidence indicates that although the young of many mammals are born with the Central Nervous System, specifically the brain, differentiated to a high degree, the human child is born with a much less differentiated nervous system. In fact, the human comes into this world with most of its cerebral differentiation yet to be accomplished. The accomplishment of this post-natal differentiation is directed along certain lines in response to experience. Thus, the human enjoys a greater freedom in matters of Central Nervous System differentiation and behavior than do other animals. As a consequence, the human can more easily participate in directing the patterns of flow of the vital force of life in the physical and psychic realms.

The Gonads and Psychic Life

Now that we have briefly considered the important contributions of the gonadal secretions to the economy of our physical life, what can we derive from this information with regard to our psychic life? We have seen that as exocrine or duct glands the gonads provide the "seeds of becoming" for the species. As a consequence, we have considered the gonads as being associated with the concept of generation or becoming. This concept of becoming is equally applicable both on the physical and psychic planes of existence.

Thus, in our psychic lives the exocrine functions of the gonads find a correspondence with the "seeds of becoming" which we generate within the depths of our being. These psychic "seeds of becoming" manifest as our aspiration and desires.

We have also seen that as endocrine or ductless glands, the gonads also secrete male and female sex hormones which exert profound influences on the differentiation of the Central Nervous System (CNS). The CNS is the instrument through whose activities we largely perceive or realize ourselves. In an analogous manner, the hidden activities associated with the process of self-generation exert profound influences on our instruments of inner perception. In other words, our aspirations and desires have a directive influence on how the soul personality perceives and responds to stimuli arising both from within and without. Such perceptions and responses are associated with the activities of the pineal gland which, as we shall see later, is one of the major centers through which bodily harmony is established and maintained.

2 The Pancreas

Through the process of energy transformation living systems evolve and grow. But one important aspect of the growth process is the closely related function of nutrition. As it relates to the growth and evolution of the human body, the function of nutrition maintains and replenishes those complex energy patterns that characterize our physiology. And since our physiology is vital to the manner in which we realize ourselves, then any organ whose functions affect the magnitude or the patterns of energy transformation in the body must, of necessity, influence behavior. The pancreas, by virtue of its intimate involvement in the process of nutrition, is just such an organ. In the human body the pancreas is located in the abdomen. See Figure 2-1.

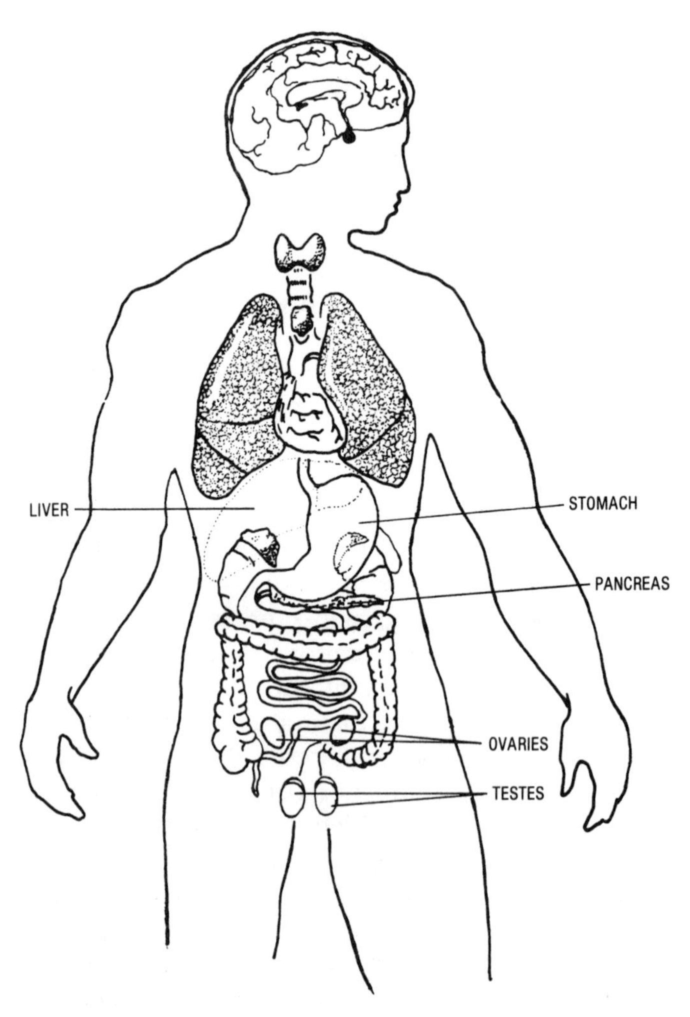

Figure 2-1. Diagram showing the location of the *Pancreas* in relation to other organs in the human body.

Exocrine Functions of the Pancreas

Like the gonads, the pancreas is dual in function, being both an exocrine and an endocrine gland. In its exocrine functions the secretions of the pancreas create the proper environment in which digestion of complex foods can occur. Accordingly, pancreatic juices, rich in sodium bicarbonate, serve to neutralize the acidity of chyme, or partially digested food, as it passes from the stomach on its way to the small intestine. Should the pancreas fail to provide this useful service, the high concentration of hydrochloric acid associated with chyme would destroy the delicate lining of the small intestine. In addition, the highly acidic chyme would inactivate the *catabolic* or breakdown enzymes which play a crucial role in the final stages of digestion. The net result would be the inability of the body to derive nourishment from food.

There is a prodigious intelligence at work in the body. Thus, the function of nutrition is no shortsighted process of "feast today and famine tomorrow." Instead, a profound wisdom dictates that nutrition be a dual process of breaking down and building up. The breakdown process involves the reduction of complex foods to simpler elements. This aspect of nutrition is accomplished by pancreatic enzymes which convert proteins to amino acids, fats to fatty acids and glycerol, and carbohydrates to sugars. On the other hand, the building up process involves the reassembly of the simpler food elements into complex storage forms. This storage aspect of nutrition is accomplished by *anabolic* or build-up enzymes in the liver. As a consequence, the body builds up stores of materials upon which it can draw in times of emergency. The major storehouse of the body is the liver.

The simpler food elements, produced through the breakdown action of pancreatic enzymes, leave the small intestine and enter the bloodstream. In the blood, these nutrients can then travel to all parts of the body. However, in order not to overtax the limited abilities of most cells to absorb large amounts of nutrients, the highly enriched blood leaving the small intestine first travels to the liver where large quantities of nutrients are removed for storage.

The Liver as Storehouse

In the liver, large quantities of glucose are removed from the blood and converted to glycogen. Glycogen is the storage form of sugar both in muscle tissue and in the liver. Glucose that has not been incorporated into glycogen is either converted to amino acids and fatty acids, or it is burnt by the liver, muscle, and other body cells to produce heat. In the burning of sugar, carbon dioxide gas and water are produced as by-products. The liver is a great source of heat in the body.

In addition to converting glucose to other storage forms, the liver is also noteworthy for its activities in assimilating amino acids into a variety of proteins. In fact, the liver is the main source of the proteins found in the blood. The liver is also the main source of vitamin B_{12}. Vitamin B_{12} prevents a type of anemia known as *pernicious anemia*, and is produced from amino acids and other simple food elements released during the digestive process. Thus, we see that many important functions of the liver are directly dependent upon the activities of the pancreas as an exocrine or duct gland.

In ancient times the liver was thought to be the seat of our emotions. Because of the profound, indirect effects of liver function on behavior, it is easy to see how such a view could have emerged. Modern research has revealed that although most cells of the body can, and do, synthesize glycogen, it is only the liver that is capable of storing it. And since glycogen is the basic substance in which the body's energy stores are safely deposited, the liver is crucial to the regulation of blood sugar levels. In its endocrine functions the pancreas regulates blood sugar levels.

Endocrine Functions of the Pancreas

Loss of the endocrine functions of the pancreas is usually associated with a condition known as *diabetes mellitus**. Under such conditions, the body and more specifically the liver, is rendered incapable of building up and maintaining vital glycogen stores. In addition, the efficiency with which the cells of the body absorb sugar, amino acids, and other nutrients from the blood is greatly reduced. And since sugar and other nutrients must first enter the cells before they can participate in the body's cycles of energy transformation, it is obvious that the endocrine functions of the pancreas are crucial to body harmony and survival. The pancreas is, therefore, a vital endocrine gland.

The pancreas produces and secretes two hormones—insulin and glucagon. These two hormones exert different but complementary influences on the energy stores of the body. Let us begin with an examination of the manner in

*Diabetes mellitus is now recognized as an auto-immune disease in which the immune system produces antibodies which destroy the insulin-producing cells of the pancreas.

which insulin exerts its overall influence in body harmony.

Insulin and Body Harmony

As indicated earlier, in the absence of insulin, the ability of cells to absorb sugar and other nutrients from the blood is greatly reduced. As a consequence of this lowered rate of nutrient removal, sugar accumulates in the blood giving rise to high levels of blood sugar. This condition is known as *hyperglycemia.* When blood sugar levels exceed a certain critical value* then the excess glucose is excreted in the urine. Sugar in the urine is one of the symptoms associated with the disorder known as *diabetes mellitus.* Because the presence of glucose in the kidney tubules prevents the reabsorption of water back into the blood from the dilute urine solution, the body loses water through excessive urination. Excessive water-loss ultimately leads to dehydration of the body and the subsequent experience of thirst.

Because in insulin insufficiency glucose and other nutrients do not enter the cells at an ade-

*160 milligrams per 100 milliliters of blood; normal blood sugar falls between 60 and 100 milligrams per 100 milliliters of blood.

quate rate, the diabetic can often suffer from a chronic experience of hunger. If the condition of insulin insufficiency is left uncorrected, then the subject indulges in excessive eating. But because the lack of insulin inhibits the entry of glucose and other nutrients into the cells of the body, regardless of the quantity of food ingested, the body loses weight and the subject experiences fatigue. In addition, accompanying disturbances in fat metabolism lead to the accumulation of certain breakdown products known as ketone bodies. Under such conditions ketone bodies act as poisons. If steps are not taken to correct the situation then the acidity of the blood increases, coma follows and, eventually death ensues.

Insulin insufficiency also has important consequences for the metabolism of the Central Nervous System. Quite apart from its influence on sugar metabolism insulin insufficiency affects the availability of certain important amino acids to the transforming and synthesizing machinery of neurons and other tissues of the body. Thus, for example, because the amino acid tyrosine is hindered in its entry into cells, its availability for the formation of thyroid hormones and the adrenal hormones, adrenalin and nor-adrenalin, is greatly reduced. Thyroid hormones are important factors in adjusting the rates at which energy is transformed in the body. On the other hand, adrenalin and nor-adrenalin are important agents insofar as the activities of the Central Nervous System and the Autonomic Nervous System are concerned.

Similarly, the relative inability of the amino acid tryptophan to enter the cells of the Central Nervous System has important consequences for the stores of the neurotransmitter, serotonin, and the closely related pineal hormone, melatonin. Thus, we see that the activities of the pancreatic hormone, insulin, have profound, indirect influences on the functioning of those structures through which we most acutely realize self.

Glucagon and Body Harmony

In contrast to the actions of insulin which favor the conversion of blood glucose to liver and muscle glycogen, the pancreatic hormone, glucagon, accelerates the breakdown of glycogen back to glucose. In this respect, glucagon acts like the adrenal hormone, adrenalin, by inducing the elevation of blood sugar levels. It should be noted, however, that although insulin acts to lower blood sugar levels while glucagon acts to elevate them, these two pancreatic hormones act cooperatively in the overall scheme of body harmony. Thus, between the lowering influence of insulin on one hand, and the elevating influence of glucagon on the other, blood sugar levels are maintained within a relatively narrow range even under conditions of great fluctuation in the amounts of ingested carbohydrates. An adequate balance between the production of insulin and that of glucagon is, therefore, essential to proper growth and evolution of the body.

Obviously then, any condition of the pancreas which results in the overproduction of insulin or the underproduction of glucagon has important consequences for the body. In either case, the body would experience low blood sugar levels. This condition of low blood sugar levels is known as *hypoglycemia*. When blood sugar levels fall below a certain critical value* then the individual behaves as though inebriated. When blood sugar levels fall much below this critical value then convulsions occur.

Pancreatic Function and the Brain

Because the brain has no glycogen stores of its own, it is totally dependent upon blood glucose for its energy supplies. When blood sugar levels fall below normal values, most people experience irritability. If the hypoglycemia is not relieved, then we may experience incoordination, emo-

*60 milligrams per 100 milliliters of blood.

tional instability, and disorientation. These symptoms disappear only when blood sugar levels are restored to normal. This restoration is ordinarily achieved through eating.

Although we tend to attribute our experience of hunger to activities of the gastrointestinal system, studies in obesity have shown that the experience of hunger has little to do with the stomach but a great deal to do with the brain. The nutrient-content of the blood is one of the signals to which certain structures of the brain are particularly sensitive. Hypoglycemia is a potent signal to centers in the hypothalamus which then respond in such a way as to elicit the experience of hunger in our consciousness.

The vital connection between blood sugar levels and some aspects of behavior is clearly demonstrated in cases involving schizophrenia and manic-depression. Studies have shown that shortly before an attack, individuals suffering from these mental disorders become increasingly anxious and restless. They quickly abandon their normally balanced diets of meat, vegetables, etc., and begin to ingest large quantities of food rich in carbohydrates. Indeed, they behave as though suffering from acute hypoglycemia. In this regard, it is interesting to note that a hormone from the hypothalamus acts directly on the pancreas to suppress the release of insulin and glucagon. This hypothalamic hormone* also acts to modulate the exocrine (digestive) functions of the pancreas. Thus, we see that the endocrine and exocrine functions of the pancreas are of such significance to physical survival that they fall under the control of certain brain centers, rather than under the control of the pituitary as do those of many other glands.

The Pancreas and Psychic Life

Now, what do the functions of the pancreas have to teach us with regard to our psychic life? Just as there are discernible correspondences between the physical functions of the gonads and analogous psychic functions, so too, there exist psychic correspondences with the physical functions of the pancreas. As with physical "seeds of becoming," our desires and aspirations must be nourished if they are to grow and evolve. Our psychic "seeds of becoming" grow by drawing unto themselves the simpler "food elements" which can be organized or assimilated into broader, more inclusive and altruistic views of life. The simpler "food elements" from which our psychic seeds draw their sustenance are pro-

*The name of this hypothalamic hormone is *somatostatin.*

vided by the concepts, notions, systems of thought, etc., which we ingest through study and individual effort. In order for psychic growth to occur, such concepts, notions, etc., must first be digested. In other words, through analysis the psychic foods that we ingest through observation and study must first be broken down into their component parts just as complex foods are first broken down into amino acids, sugars, etc., by the catabolic enzymes of the pancreas. And just as the synthesizing enzymes of the liver reassemble the simpler food elements into proteins, vitamins, etc., so too simple ideas, thoughts, etc., are reassembled or assimilated through the psychic process of synthesis into patterns of reality harmonious with the growth and evolution of our desires and aspirations.

Nevertheless, it is important to bear in mind that even the psyche must be provided with a balanced diet from which to build the type of character we wish to develop. Accordingly, we should exercise responsible choice in assuring that we do not overindulge in mental or other activities which, like the cravings of the unfortunate schizophrenic or manic-depressive, could lead to severe imbalances in our higher psychic functions.

Just as the growing and evolving body can draw only on those materials that have been previously ingested, digested, and stored, so too, in psychic growth and evolution, our psychic seeds can grow only by drawing upon material which the personality has previously ingested, digested, and stored. Psychic growth is limited only by experience. If we neglect to accumulate psychic experience through contemplation and meditation, then our "seeds of becoming" are likely to waste away and die just as does the body whose liver does not accumulate energy stores because of an imbalance in the production and secretion of the pancreatic hormones insulin and glucagon. The pancreas, in its dual function as an exocrine and endocrine gland, therefore, has much to teach us in regard to the judicious storage and dispensation of our psychic energy which flows naturally along the channels of our thoughts and experiences.

3 The Adrenals

Anyone who has ever witnessed the behavior of animals in their natural environments cannot help but be impressed by the difference between the pugnaciousness of the lion, on the one hand, and the timidity of the fleeting rabbit on the other. Such conspicuous differences in animal behaviors are related, in large measure, to the functions of the adrenal glands as they manifest the energies associated with self-assertiveness or self-expression.

As we saw in our earlier considerations of the gonads and the pancreas, the Law of Duality manifests as the endocrine and exocrine functions of these two glands. However, in the case of the adrenals, this great law can be seen to manifest not only at a functional level but also at the level of the origin and structure of the glands. The adrenal glands are two in number, one on the top of each kidney. See Figure 3-1.

Dual Structure of the Adrenals

The adrenals or suprarenal* glands are constituted of two types of tissue. The cortex, or outer portion of the glands, arises from the same type of embryonic tissues as do the gonads. On the other hand, the medulla, or the inner portion of the adrenals, arises from the same embryonic tissue as does the Autonomic Nervous System. Accordingly, the hormones of the adrenal cortex are steroids, belonging to the same class of molecules as do the hormones of the gonads, while the hormones of the medulla belong to a class known as neurotransmitters.

From a developmental point of view, the two tissue-types of the adrenal glands have not always been physically associated in the form of one organ or gland. Thus, in lower vertebrates the "cortex-to-be" and the "medulla-to-be" are quite separate. At that level of development represented by the reptiles, the two tissues are adjacent to each other, while at the level of birds,

*The adrenals are also called suprarenal glands because they are located atop (supra) the kidneys (renal).

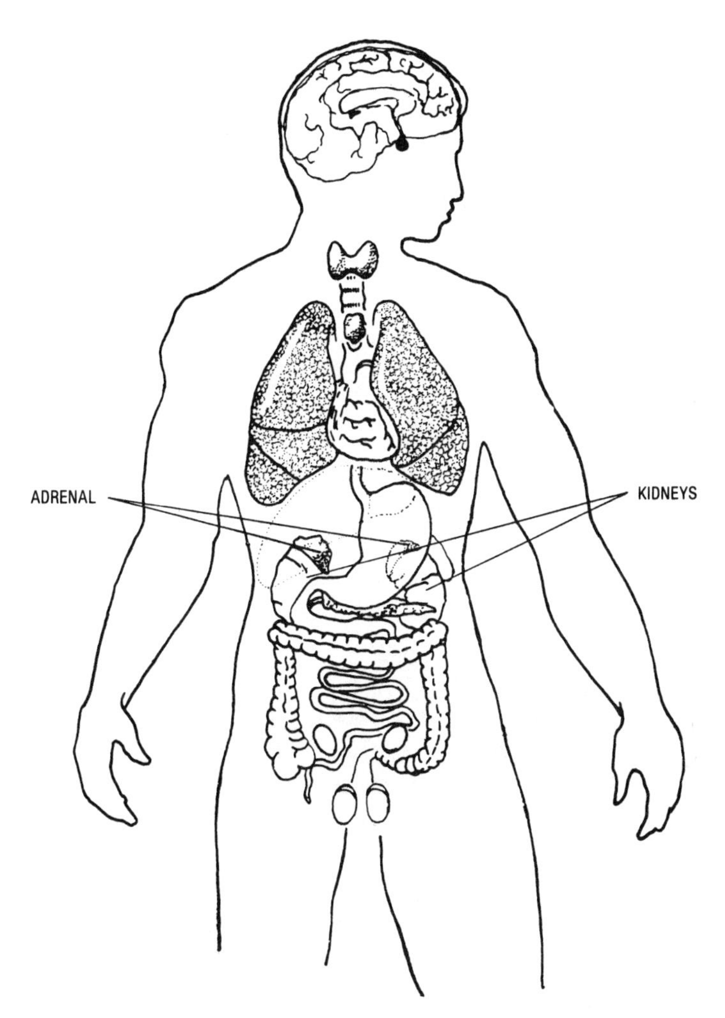

Figure 3-1. Diagram showing the location of the *Adrenal Glands* in relation to other organs of the human body.

the two come together for the first time. Thus, at the human level we find a gland in which two discrete functions are tightly integrated.

Adrenal Cortex Size and Behavior

Several studies have shown that the size of the adrenal cortex seems to vary directly with the pugnaciousness or self-assertiveness of an animal. Thus, in charging, fighting animals, such as lions and tigers, the adrenal cortex is remarkably wide. On the other hand, timid, fleeting animals, such as the rabbit, are noted for their conspicuously narrow cortexes. This relationship between cortical width and self-assertive aspects of animal behavior is further demonstrated by the fact that wild members of a given animal species have wider adrenal cortexes than do their domesticated brothers. Interestingly, human beings possess a cortex relatively larger than that of any other animal species. The relative size of the adrenal cortex may therefore be intimately associated with those qualities of being that we recognize as human.

In some subtle way, human brain development is associated with the functions of the adrenal cortex. This fact is demonstrated by the following observations. Between the fourth and sixth weeks following conception, the adrenal glands of the developing fetus are twice as large as the kidneys. Most of this relatively large size is contributed by enlargement of the cortex. The human is the only species in which this predominance of cortex size has been noted. Should this predominance of the cortex over the medulla not occur, i.e., if the proportion of cortex size to the medulla size remains low, as it does in other species, the brain does not develop properly. In fact, the development of a grossly malformed creature is often the result. The development of the human brain, with its number and complexity of cells, may therefore be controlled by the functions of the adrenal cortex.

Structure and Functions of Adrenal Cortex

The adrenal cortex consists of three well-defined layers*. Each of these layers produces and secretes its own complement of steroid hormones. Thus, the outermost layer produces a hormone which functions primarily in controlling mineral metabolism. By virtue of its effects on the ionic content of the blood, this hormone

*The three layers of the adrenal cortex are: (1) The outermost Zona Glomerulosa which produces the mineral-controlling hormone, aldosterone; (2) the central Zona Faciculata which produces hydrocortisone, cortisone and cortisosterone; and (3) the innermost Zona Reticularis which produces sex hormones.

also exerts profound influences on the composition of the blood and, therefore, on water balance, blood volume, and blood pressure.

The center layer of the adrenal cortex produces a class of steroid hormones which control the conversion of protein to carbohydrate. As may be expected, the actions of these hormones have profound consequences for liver-glycogen stores, as well as for blood sugar levels. In addition, the hormones of this central layer of the adrenal cortex are intimately associated with the body's reactions to stress such as exercise, trauma, burns, and infections. In fact, these hormones function as anti-inflammatory and antiallergic agents by mobilizing certain cell types produced by the thymus gland.

The innermost layer of the adrenal cortex produces sex hormones mainly of the androgenic or male type, and therefore, has an influence on fat metabolism. In recent years, a derivative of one of these male hormones* has been shown to exert profound antiobesity effects by inhibiting some aspects of glucose metabolism. Interestingly, because of its action as an antiobesity agent, this

*The name of this hormone is Dehydro-Epiandrosterone, abbreviated DHEA.

hormone also appears to be a potent antagonist to tumor growth and development. Other androgenic hormones of adrenal origin are associated with the masculinization of the Nervous Systems which occurs in female embryos exposed to the influences of hyperactive adrenal glands.

Individual Sensitivity and Activity of Adrenal Cortex

Not surprisingly, hypo-activity of the adrenal cortex has profound influences on certain aspects of behavior. A case in point relates to the extreme hypersensitivity of subjects suffering from Addison's Disease, a disorder generally characterized by an insufficiency in function of the center layer of the adrenal cortex. These subjects demonstrate a taste sensitivity some 150 times more acute than do normal individuals. Even more startling are the observations made with respect to the sense of smell. For example, even though salt has little or no smell to most people, it contains enough chlorine gas to be easily detectable by patients with Addison's Disease. Treatment with steroid hormones characteristic of the center layer of the adrenal cortex quickly abolishes these hypersensitivities and returns the acuity of the senses to the normal range. Similar observations have been made with respect to the sense of hearing. Interestingly, subjects with Cushing's Disease, a disorder, generally characterized by hyperactivity of the center layer of the adrenal cortex, are relatively insensitive to tastes and smells.

In contrast to the observations made with regard to hypersensitivity of the senses of smell, taste, and hearing, insufficiency of hormones from the adrenal cortex is also associated with deficiencies in the ability to integrate and interpret sensory inputs. Accordingly, although subjects with Addison's Disease are exquisitely sensitive to sounds, they show a surprising lack in their ability to discriminate between a steady and warbling sound, for example, or even to determine the direction of the source of sounds. Again, their performance can be made to return to normal simply through treatment with steroid hormones from the central layer of the adrenal cortex.

Cortical Hormones and the Nervous System

The discrepancy between sensitivity and interpretation is due to the fact that the type of adrenal insufficiency characteristic of Addison's Disease is associated with a more rapid conductance of impulses within nerve cells, but a much reduced rate of transmission of impulses between nerve cells. It appears, therefore, that the hormones of the central layer of the adrenal cortex affect the nervous system in such a way as to reduce our sensitivity to stimuli and, at the same time, to increase our ability to interpret and to integrate sensory inputs. It is noteworthy, therefore, that our sensory acuity appears to reach a maximum level at precisely the same time that the blood levels of steroid hormones from the central layer of the adrenal cortex are at their minimum. Thus, for the normal individual who goes to bed between 10:00 and 11:00 p.m., this period of minimum hormonal levels and maximum sensory acuity occurs at about 3:00 a.m., a time known to Rosicrucians as being particularly appropriate for the reception of impressions from the subconscious mind.

The Adrenal Medulla and Emotions

Unlike the hormones of the adrenal cortex, the influences of which we are usually quite unconscious, we do not often remain unaware of the impact of the adrenal medullary hormones on the body mechanisms. The reasons for this difference in perception can be attributed to the fact that the adrenal medulla is closely associated with those structures through which we are most acutely aware of self. When we come to realize that the adrenal medulla is itself a collection of nerve cells belonging to the sympathetic arm of the Autonomic Nervous System, we can more readily appreciate the significance of this difference. In fact, any sympathetic stimulation, be it an experience of fear, anger, pain, joy, ecstasy, or love, will trigger the release of the hormones from the adrenal medulla, adrenalin and nor-adrenalin. Not surprisingly, the adrenal medulla has often been referred to as "the mirror of our emotions."

In addition to triggering the release of adrenalin and nor-adrenalin, emotional stresses also cause the shunting* of blood away from structures such as the skin, liver, spleen and gastrointestinal tract. This shunted blood is rerouted toward the muscles and the brain where it is most needed. At the same time, adrenalin stimulates the release of liver glycogen to the cells and causes the heart to beat faster and stronger, thereby enabling the blood, laden with important nutrients, to get more rapidly to the vital areas. Simultaneously, the rate of breathing is increased in order to insure proper vitalization of the blood as it passes through the lungs. Thus, the adrenal medulla easily qualifies as the gland of energy mobilization. What then is the relationship between the functions of the adrenal cortex

*This shunting of blood away from the liver, spleen, pancreas, gastrointestinal tract, etc., is mediated through the action of the solar plexus.

and those of the adrenal medulla? In answer to this question, let us briefly review the sequence of events which ensues when we call the functions of adrenals into action by the use of strong emotions.

When the interpretation of our perceptions demand action, be it defensive or offensive, our emotions are engaged. Depending upon our past experiences, our emotional responses may manifest as fear, anger, courage, or ecstasy. These reactions occur in the higher brain centers which are in direct communication with the Autonomic Nervous System (ANS). When our emotional responses are those of fear, anger, etc., the sympathetic or arousing arm of the ANS is called into action. As a consequence, sympathetic nerve impulses travel to the adrenal medulla and trigger the release of adrenalin into the bloodstream.

Adrenal Hormones and Body Reactions

Adrenalin in the bloodstream soon reaches the heart which, in response, beats harder and faster sending blood quickly to the brain. In the brain, adrenalin acts upon the hypothalamus which, in turn, causes the anterior pituitary gland to release a hormone* which acts upon the adrenal

*The name of this anterior pituitary hormone is Adreno-Cortico-Tropic Hormone—ACTH.

cortex. In response to the actions of the cortex stimulating hormone from the pituitary, the adrenal cortex releases steroid hormones from its central and innermost layers. These hormones then produce their characteristic influences in preparing the body to meet the demands of the stressful situation.

Thus, we see the high degree of cooperation that exists between the dual structures which comprise the adrenal glands. As a whole, the adrenal glands are, therefore, the glands of stress and strain or more specifically the glands of mobilization. The purpose which underlies any adrenal mobilization is defined by the interplay between "heart and head." This aspect will be dealt with in our discussions of the thymus and the pineal glands.

In light of the foregoing, it is clear that removal of adrenal functions has severe consequences for the survival of the body. Because of their influence on the chemistry of the blood, the functions of the adrenal glands are intimately associated with the electrical conductivity and electromagnetic properties of the body. And because we exist in an ocean of electromagnetic waves, the adrenal glands have much to do with the quality of our psychic attunement with our environment.

Qualitative chemical analysis has indicated that through fear and anger, it is possible to exceed or exhaust the capabilities of the adrenals to meet the demand for hormones. Such exhaustion often disengages us from many beneficial influences inherent in the electromagnetic aspects of our environment, and may cause us to experience fatigability, increased sensitivity to cold, loss of zest, indecision, irritability, and depression. Interestingly, some modern scientists contend that many of the diseases that plague our highly industrialized societies have their origins in adrenal exhaustion. It seems clear then, that as individuals we would do well to adjust our emotional responses to life's experiences so as to minimize the risks of adrenal exhaustion. But how do we learn to master such adjustments?

Adrenals and Psychic Life

Just as our emotions trigger the release of adrenal hormones which mobilize or propel our physical being into action, so too, our emotions mobilize our psychic being along the lines determined by our desires and aspirations. As a consequence, the degree of harmony that we experience in our daily lives is related to that which we desire and that to which we aspire. Naturally, then, we would wish to ascertain that

that which we desire is in harmony with our highest ideals.

Careful observation shows that the negative emotions of fear and anger often arise within our consciousness because of misunderstanding, or as a result of a misalignment between our immediate desires and those ideals which we nurture deep within our heart of hearts. Such lack of understanding can be regarded as a sort of "psychic indigestion" which arises as a result of failure to analyze or digest properly some aspect of a previous experience. In this vein, the personality may be said to have derived little "nourishment" or meaning from the experience in question. Accordingly, we often expend great quantities of psychic and physical energy on matters not in keeping with our "highest good." The result is that we too often leave ourselves so exhausted that we are unable to attain our most cherished goals. Such is often the situation which typifies a conflict between "head and heart." In this regard, Rosicrucians know well the importance of contemplation and meditation.

4 The Thymus

"*A young boy* is admitted to the hospital in a coma. His condition has been brought on by a severe viral pneumonia and is accompanied by a high temperature and an inability to breathe unassisted. Everything possible is done to resuscitate the boy, but to no avail. It is feared that, barring a miracle, he will die.

"Blood tests show that the count of a certain type of blood white cell is only one-fifth the normal level. Such a finding is indicative of a low thymus activity. As a consequence, the patient is given an injection of thymus extract. Within twenty-four hours, the dismal clinical situation is completely reversed. The boy has regained consciousness, his temperature is normal, and he is breathing unassisted."

The preceding account is an excerpt from an actual medical case history and dramatically illustrates the miracle inherent in the endocrine activities of the thymus gland. And yet, the thymus, for all of its miraculous influences, is not a vital endocrine gland. That this is so is illustrated by the fact that, although abnormal in many ways, animals and humans alike are capable of existing without the thymus gland. What then are the endocrine functions of the thymus?

Endocrine Functions of the Thymus

Physiologically, the thymus has two major functions. These functions are growth promotion and immunological competence. Very early in the life of the developing fetus, the thymus becomes a focus of recognition whereby the budding lymphatic system is harmonized with the individuality of the developing body. In this sense, the thymus is at the very heart of the body's sense of self. In fact, it may be said that the thymus becomes the focal point through which the developing fetus is harmonized in all of its parts. This harmonizing function is reflected in the activities of a thymic hormone known as homeostatic thymic hormone or HTH for short. HTH appears to be responsible for harmonizing the entire glandular system and may, therefore, be instrumental in determining the "personality" of the endocrine system of psychic centers. The traditional importance attached to the thymus gland and its attraction for the psychic qualities of the personality manifesting through the body may, therefore, find scientific validation.

The thymus is now well recognized as the gland of childhood. Anatomically, the thymus is a pyramid-shaped organ which, in the human, is located immediately beneath the breastbone at the level of the heart. Refer to Figure 4-1. During the childhood period of life, it may extend from the neck region all the way down to the diaphragm in the region of the abdomen. Following puberty, the thymus diminishes in size so that during the adult period of life it may be considerably smaller than it was prior to puberty. Refer to Figure 4-2. Interestingly, absence of the thymus in the human is often associated with mental retardation and a greatly reduced pineal gland. As we shall see later in this volume, the pineal exerts important influences on the rate at which the personality unfolds in the body.

The influence of the thymus in harmonizing the body's sense of self finds expression in the fact that during fetal life, as well as during the period prior to puberty, the thymus, in conjunction with the pituitary gland, coordinates harmonious body growth. In fact, the growth-

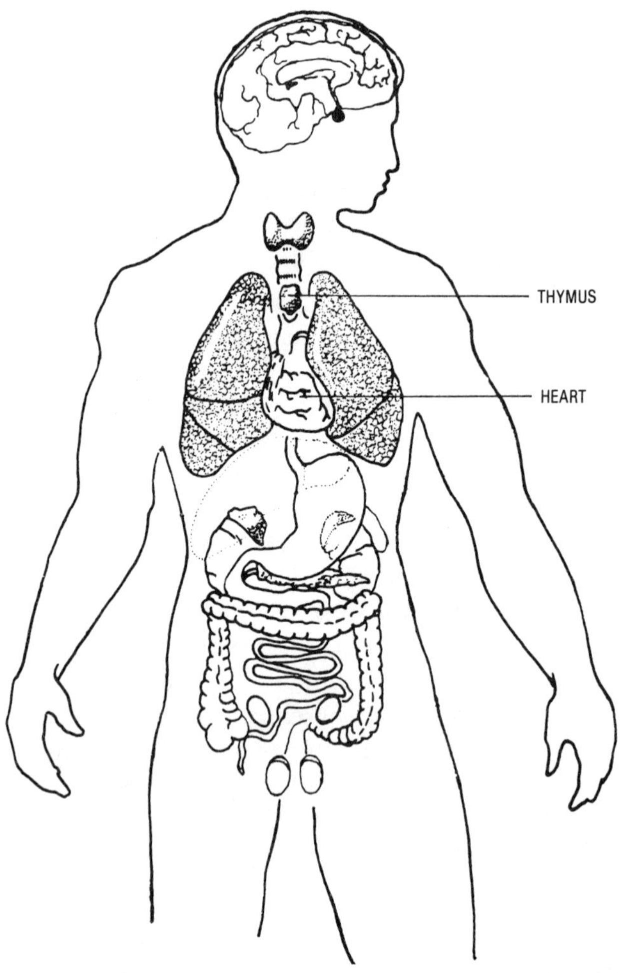

Figure 4-1. Diagram showing the location of the *Thymus Gland* in relation to other organs of the average adult human body.

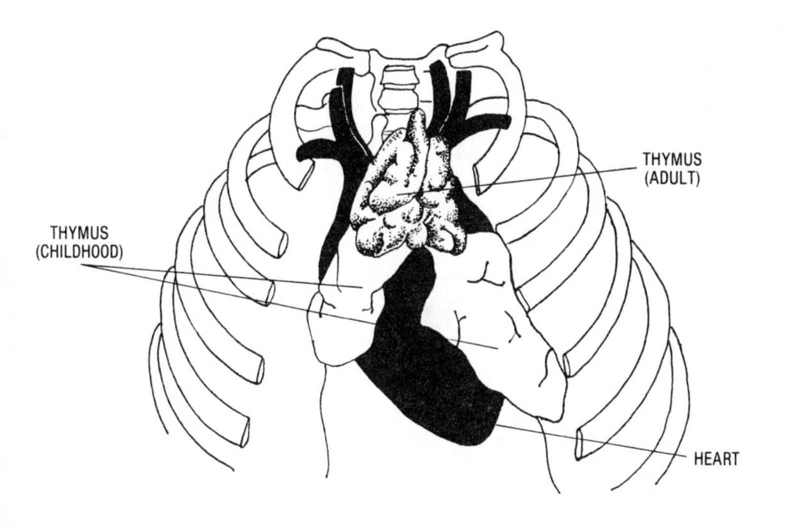

Figure 4-2. Diagram showing relative sizes of the *Thymus Gland* during childhood and during adult life.

promoting activities of the pituitary are, themselves, dependent upon a normally functioning thymus gland. Prior to puberty, the thymus also exerts important regulatory influences on the activities of the thyroid, the adrenals, the pancreas and the gonads. This is accomplished by means of control of the synthesis of many important hormones of the pituitary gland. The thymus, therefore, plays an important role in determining the rate of unfoldment of the body. Following puberty, however, the thymus relinquishes many of its regulatory activities and becomes principally concerned with maintaining body integrity through its influence on the lymphatic system.

As though in recognition of its early association with the parathyroid glands, the thymus produces a hormone which contributes to the control of blood calcium levels. This calcium lowering component of the thymus appears, in some obscure way, to be involved in the regulation of calcium metabolism. Calcium, as we shall see when we consider the parathyroids, is an important factor in the transmission of nerve impulses. In this regard, it is of interest to note that another thymic hormone, thymin by name, is involved in the transmission of nerve impulses to muscle fibers. Thus, in addition to its other functions, the thymus also has important influences on the psychic sense of self, particularly as this sense manifests through the activities of the nervous system. It may be more than mere coincidence then that the thymus is more prominent in the more self-assertive animals, such as the carnivores, than it is in the more timid herbivores.

The Thymus and the Body's Sense of Self

In its capacity as the "heart" of the body's defense mechanisms, the thymus programs a large number of cells specialized in body defense. These cells belong to a general class of lymphocytes or white cells found in the blood and lymph. Lymphocytes respond to such inharmonious conditions as bacterial invasion, infection, injury, stress and/or foreign materials. Although not yet identified, a hormone may be released by the adult thymus in response to conditions of stress such as fright, shock, etc. In this regard, it is worthwhile recalling that when the interpretation of our perceptions demands action, be it defensive or offensive, the functions of the adrenal gland are set into motion. Part of the adrenal response to such stresses involves the secretion of a steroid hormone which mobilizes the expulsion of certain cells from the thymus. These cells are known as thymus-derived lymphocytes or T-lymphocytes. Beginning in early life, T-lymphocytes circulate in the body and provide the earliest physical defense mechanisms. T-lymphocytes are not only expelled from the thymus by steroid hormones, but a large number of these cells are destroyed by certain of the steroid hormones. Thus, we see that the "negative" emotions such as fear and anger can greatly reduce the body's ability to defend itself against invasions by foreign materials.

Traditionally, the thymus is recognized as being associated with the individual's first experience of "self." It, therefore, seems more than coincidental that at the cellular level, thymus-derived lymphocytes should be capable of distinguishing between those cells which harmonize with the physical "self" and those that interfere with body harmony. Today, questions concerning *how* and *why* thymus-derived lymphocytes "know" the difference between "self" and "non-self" reflect the more traditional concerns with regard to the personal sense of "self" versus "non-self" enjoyed by every conscious human being.

The role of the thymus as the "heart" of the body's sense of "self" is mediated through the activities of its T-lymphocytes. These lymphocytes fall into two classes identified as T_1 and

T_2. The T_1-lymphocytes are concerned with the regulation of the body's defense mechanisms, being involved in such immunological reactions as the production of antibodies and the rejection of transplanted organs. On the other hand, T_2-lymphocytes are responsible for the actual carryingout of those cellular reactions that characterize allergic responses, the destruction and rejection of foreign grafts, etc.

Stress and the Thymus

In recent times, the role of stress in the genesis of many modern diseases is receiving increasing attention. Thus, it has been determined that the circulating levels of T_1-lymphocytes are greatly reduced under conditions of stress. This relationship between stress and the depletion of T_1-lymphocytes, takes on added significance when it is recalled that certain steroid hormones from the adrenal cortex and the gonads cause the release destruction of lymphocytes by the thymus. Many of these lymphocyte-releasing steroids are among those secreted by the adrenals in response to stress.

The significance of the relationship between the interpretation of our perceptions, stress, and the preparedness of the body to defend itself against invasion, is dramatically underscored by the results of an experiment conducted at Brown

University in Rhode Island during the mid 1970's. A number of undergraduate students were subjected to a set of conditions which could be interpreted as being stressful. Psychological testing allowed separation of the students into two sub-groups, one which considered the conditions stressful and the other which did not. A blood sample was taken from each student following exposure to the test conditions. From each blood sample the white cells or lymphocytes, were separated and treated in a manner which would challenge them to produce antibodies*. The lymphocytes from those students who interpreted the test conditions as being stressful were by far less able to produce antibodies than were the lymphocytes from those students who interpreted the test conditions as being non-stressful. In other words, the sense of inadequacy experienced by the stressed students seemed, by some unknown mechanism, to find expression as the inadequacy of their lymphocytes to meet the challenge of adequate antibody production. Such experimental results give us ample reason to reflect upon the possible influences of our

*Antibodies are large protein molecules produced by lymphocytes to inactivate invading foreign materials. T-lymphocytes participate in this reaction.

internal reactions on the body's defense capabilities. In this regard, it is noteworthy that modern researchers are finding that certain diseases are more frequently associated with certain personality-types than with others. In fact, according to some researchers, as many as 85-90% of all physical disorders have some significant psychic contribution.

The Thymus and Psychic Life

Just as during childhood, the thymus harmonizes and coordinates the unfoldment of the physical body in accordance with the potential and possibilities inherent in the environment, so too, our sense of self harmonizes and coordinates the unfoldment of the potential inherent within Self. But of what does the "psychic self" consist? Just as the potential physical body represents the unfoldment of the seeds-of-becoming provided by the gonads, so too, our psychic self represents the unfoldment of the seeds-of-becoming spawned by our deepest desires. And just as the physical body is subject to certain limitations imposed by its genetic makeup, so too, the psychic self is subject to certain limitations imposed by the meaning which we have assigned to our past experiences. However, unlike the physical body, we are capable of transmuting the psychic self.

Most of us recognize and accept the idea that the limitations of the physical body are imposed by powers beyond our ability to challenge. For most of us, the interaction between the limited "self" and the limitless possibilities of the "non-self," is quite automatic, leaving little room for personal input. Such a view is, itself, a limitation but it is not one imposed by powers beyond our ability to challenge. In fact, such a limitation is entirely self-imposed and is, therefore, amenable to change.

Desire is the means by which we may institute change in our habitual ways of responding. When we are successful in imposing such changes, we are said to have changed our reality or the way in which we realize our world. As a consequence, we experience a change in our sense of self which, in turn, brings about corresponding changes in our psychic centers and mental attitudes. Major changes along these lines are often referred to as self-initiation and manifest the alchemical process of evolution. In this regard, it may be meaningful to recall the old saying: "As a man thinketh in his heart, even so is he."

5 The Thyroid and Parathyroids

Of all the endocrine glands, the thyroid probably qualifies as the first to have captured man's attention. Under certain conditions this gland can become so greatly enlarged that it is difficult not to notice it. Such enlargements are characteristic of the glandular conditions known as goiter.

During the 1800's, goiter was well known to be associated with specific geographic locations in Europe and the United States. In England for example, the disorder was so common around Derbyshire that goiter became known as "Derbyshire Neck." Similarly, in one small village in Switzerland during 1848, of a total population of only 1,472, there were 900 cases of goiter plus 109 cases of cretinism. A brief sixteen years later, following a change in water supply, there were only 39 cases of goiter and 58 cases of cretinism. What could there have been in the water that could cause such a dramatic reversal?

It had been known that thyroid secretions contain a certain quantity of the chemical element known as iodine. It was also known that in certain districts the development of goiter had been prevented by adding small quantities of iodine to the food and water supply of the region. For these and other reasons it is now accepted that goiter is primarily an iodine-deficiency disorder in which the thyroid gland enlarges in an effort to extract and store the minutest traces of iodine introduced to the body.

Of all organs in the body, the thyroid gland has an unusual affinity for iodine. Iodine taken in the diet is quickly absorbed by the thyroid gland and incorporated into the thyroid hormones known as *thyroxine* and *tri-iodo-thyronine,* abbreviated as T_4 and T_3 respectively. These two hormones are secreted into the bloodstream which then carries them to all tissues of the body. But what are the functions of these thyroid hormones?

Thyroid Hormones and Animal Transformations

Anyone who has ever observed tadpoles over prolonged periods of time cannot escape being impressed by the wonderful transformations which bring forth air-breathing frogs from these water-breathing, fishlike little creatures. This wonderful metamorphosis from tadpole to frog takes place over a period of a few days or a few weeks depending upon the temperature of the environment, and involves more than just the obvious. For example, in addition to the obvious external changes, there are important, but not so obvious inner changes. Among the more easily recognized of these inner changes which occur during the transformation, is the reduction of the several feet long intestine of the tadpole to a mere couple of inches in the frog. In addition, following metamorphosis, physiological changes are such that the "pace of life" within the body of the frog is much greater than it was in the tadpole. This acceleration in the "pace of life" is associated with a greater access to, and enhanced utilization of, oxygen gas. If for one reason or another the tadpole is deficient in thyroid function, the wonderful transformations which underlie the transition from a water-breathing creature to an air-breathing frog would not occur. On the other hand, if the tadpole were to be exposed to thyroid hormones too early in its development, then the water-to-air transition would occur prematurely and a very tiny frog would be the result. The functions of the thyroid gland are seen, therefore, to be intimately associated with transformation or transition from one level of organization to another.

The Thyroid in Man

In the human, the thyroid gland is a shield-shaped organ located in the neck at about the level of the voice box or larynx. See Figure 5-1. According to some scientists, the thyroid was once a sex gland, having been previously associated with the ducts of the sex organs or gonads. Although in higher animals there are no known physical connections between the thyroid and the gonads, important physiological associations are known to exist. Thus, proper thyroid function is necessary for development of the reproductive mechanisms in humans and other mammals. In the absence of thyroid function the child, like the tadpole, does not undergo the transition into adulthood.

Just as the transition from tadpole to frog is accompanied by important inner changes of a physical and physiological nature, so too puberty in the human child is accompanied by important inner changes. It is not surprising then that the inability of the child to undergo puberty in the absence of thyroid functions also has important consequences for the development of mental maturity. For example, lack of thyroid function

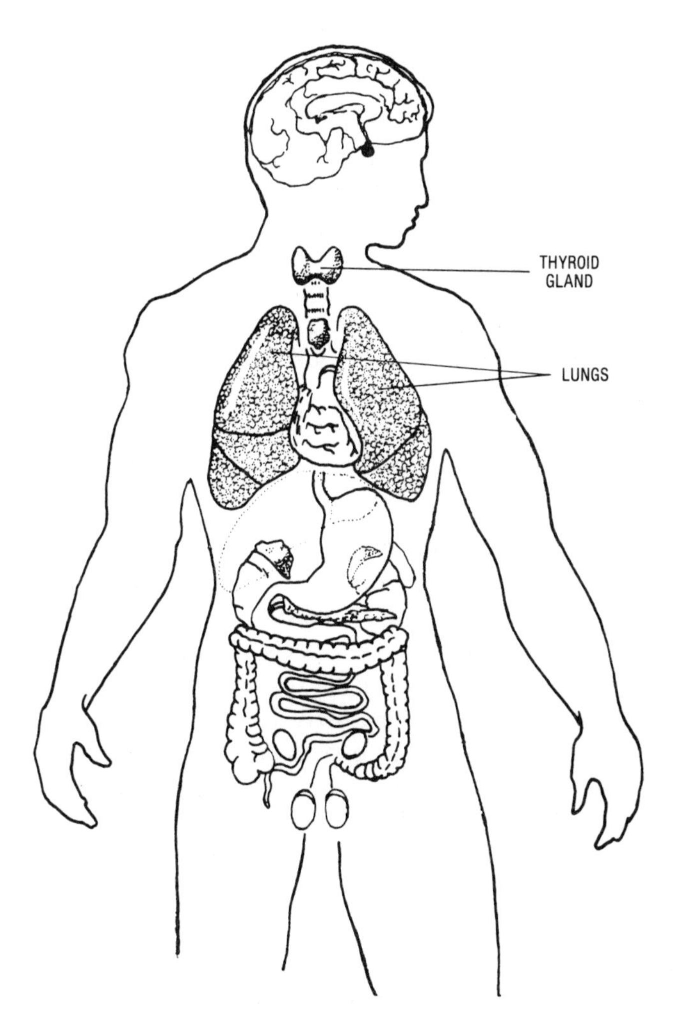

Figure 5-1. Diagram showing the location of the *Thyroid Gland* in the human body.

during the childhood period of life leads to a state of mental dullness known as cretinism. Interestingly, lack of thyroid function during adult life also results in mental dullness, in addition to apathy, drowsiness, and sensitivity to cold. This condition is known as myxedema and, like cretinism, can be ameliorated through treatment with thyroid hormones. Thus, in the human at least, thyroid function is not only associated with the development of mental alertness but is also necessary for maintaining that alertness or degree of self-awareness.

The Thyroid and Oxygen Utilization

As indicated earlier, the acceleration in the pace-of-life which accompanies the transition from tadpole to frog is associated with a greater access to, and enhanced utilization of, oxygen gas. All life forms on Earth are dependent upon oxygen gas in order to persist in their sense of being. In animal systems, thyroid hormones regulate the rate at which oxygen is utilized by the tissues and organs of the body. Further, in the absence of thyroid function the rate at which energy is transduced or transformed in the body is reduced by 30-40%. Energy transduction in the body is firmly linked to oxygen utilization.

The requirement for oxygen by all life forms is related to the electromagnetic character of the oxygen atom. Oxygen atoms have a great affinity for negative electrons. The magnitude of this affinity for electrons is such that in living tissues oxygen atoms attract electrons away from the sugars and other types of molecules which serve as fuels for driving the machinery of the body's metabolic processes. Under the conditions which normally prevail in the body, electrons, in being attracted and trapped by oxygen atoms, must release large quantities of energy. This released energy is electromagnetic in nature, some of it manifesting as vibrations of heat, some as light, and some as other vibrations of such high frequency that we may refer to them only as "auric." Thus, as electrons move down energy gradients to become trapped by oxygen atoms they provide the energies of life and, in addition, contribute to that electromagnetic condition or aura which surrounds the body.

From the foregoing, it seems clear that thyroid hormones, by some unknown mechanism, regulate the rates at which electrons move down energy gradients to become trapped by the oxygen atoms which go into the formation of carbon dioxide and water molecules, ultimate products of tissue metabolism. The influence of thyroid hormones in regulating the rates of oxygen consumption and electron flow in the body is such that many persons with hyperactive thyroid glands become so electrified as to administer unintentional electric shocks to anyone who touches them. In addition, during the early 1980's, the thyroid hormone T_3, has been linked to premature death in infants—a syndrome known as "crib-death."

In animal systems, adequate oxygen utilization is necessary for the maintenance of self-awareness. It seems clear then that the thyroid gland functions to maintain adequate responsiveness to the environment by regulating the rates of energy transformation within the body. However, it is important to recognize that although the thyroid gland regulates the rate of energy transduction within the body, it does not itself regulate body temperature. The actual body temperature of any system represents a balance between the rate at which heat energy is produced within the body and the rate at which heat energy is lost to the surrounding environment. In the human system, body temperature is regulated by a brain structure known as the hypothalamus.

Thyroid Function and the Nervous System

Activity in the nervous system is influenced by thyroid hormones. This influence is exerted both at a physical level and physiological level. Thus, under conditions of inadequate thyroid function, nerve cell connections in the brain and other parts of the nervous system are defective. Such defects in the physical integrity of the nervous system render adequate responsiveness to the environment impossible. In addition, there is evidence to the effect that during the embryonic period of development the activities of the thyroid gland in some way determine the etchings according to which the different organs will develop. In a sense, then, the thyroid determines the characteristics of the physical vehicle by means of which one realizes self.

The influence of the thyroid on the physiology of the nervous system is most readily seen in its effects on metabolism. The nervous system, particularly the brain, utilizes the sugar, glucose, as its principal energy source. Energy is most efficiently derived from glucose when the sugar is reduced ultimately to carbon dioxide and water. And since this reduction is totally dependent upon oxygen utilization, the importance of adequate thyroid function for proper activity of the nervous system becomes obvious. Little wonder, then, that individuals with subnormal thyroid function manifest mental dullness, while those with above-normal thyroid function manifest nervousness, irritability, overexcitability and overly rapid responses.

Thus, we are now better able to appreciate statements to the effect that without the thyroid gland there can be no complexity of thought, no learning, no education, no habit-formation and no responsive energies with which to meet the challenges of life. Without the thyroid there could be no physical unfolding of the faculty and function of reproduction of any kind—neither physical nor psychic. The thyroid appears to be essential to the laying-down of memory and therefore regulates the interchange or flow of energy between past and present. Because thyroid removal without replacement therapy leads to degeneration of nerve cells and their network of connections, a properly functioning thyroid is responsible for maintaining associative memory. Furthermore, a properly functioning thyroid determines the rate at which we reason, judge, and perceive; it controls the rate at which we realize our inner reactions and impressions. In short, the thyroid gland not only regulates the rate at which impressions arising from internal and external sources register in objective consciousness, but also the rate at which meaning is assigned to present experiences in relation to the past.

The Parathyroids

In the human, the parathyroid glands are four in number and are embedded in the body of the thyroid gland. See Figure 5-2. Like the thyroid, the parathyroids are concerned with metabolism. However, the parathyroids are concerned exclusively with mineral metabolism while the thyroid is primarily concerned with energy transduction. In this regard the parathyroids contribute to the amounts of the minerals calcium and phosphorus found in bone. The parathyroid hormone, known as parathormone or PTH, in causing demineralization of bone also contributes to raising the level of calcium in the blood. On the other hand, a thyroid hormone, known as thyrocalcitonin, causes blood calcium and phosphorus to be incorporated into bone and thus contributes to lowering the level of calcium in the blood. In other words, as far as calcium metabolism is concerned, the thyroid and parathyroid glands work in opposition to each other. In this "push-pull" manner both the thyroid and the parathyroids function in maintaining adequate blood levels of calcium.

The Parathyroids and Calicum Utilization

Absence of the parathyroids results in a pronounced fall in the level of calcium in the blood and brain. This decrease in calcium is accompanied by an increase in blood phosphorus *plus* an astounding increase in the excitability of nerves. Nerve excitability has been found to

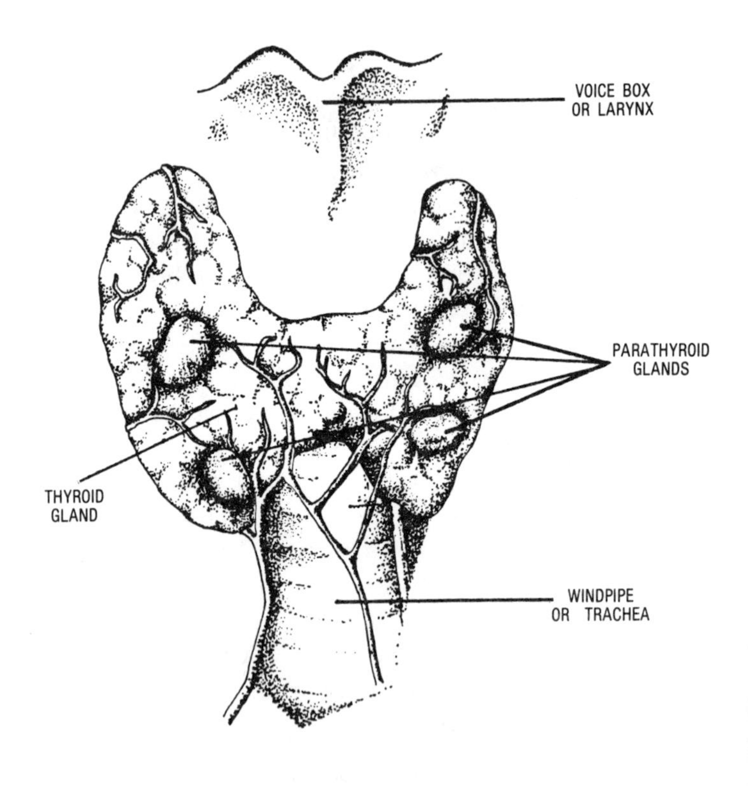

Figure 5-2. Diagram showing the location of the *Parathyroid Glands* in relation to the *Thyroid.*

increase some 500-1000 times in the absence of parathyroid function. Such levels of nervous excitability can cause an animal to go into convulsions in response to such simple acts as a sudden noise or allowing light into its darkened environment. In man, a condition of nervous excitability associated with grossly subnormal parathyroid function is known as tetany. Tetany, if left untreated, can lead to death. The parathyroids are, therefore, vital endocrine organs.

Administration of parathyroid hormone to animals deficient in parathyroid function leads to increased tissue and blood content of calcium and a marked decrease in the electrical excitability of nerves. Parathyroid hormone also indirectly regulates the excitability of other organs such as the heart, stomach, and intestines and, therefore, controls the responsiveness of the entire organism to the rhythms of the Cosmic.

The Parathyroids and Excitability of the Nervous System

Children with poor parathyroid function usually have poor teeth and are hypersensitive. Because of their hypersensitivity such children develop protective behavioral responses and other psychological devices which are sometimes diagnosed as peculiar complexes, phobias, etc. On the other hand, hyperactive parathyroids may lead to an underexcitable nervous system and hence to mental sluggishness. Adequately sensitive nervous systems—both the Central Nervous System and the Autonomic Nervous System—are important requirements for each student on the Path.

During the latter half of the childhood period, children with subnormal parathyroid function often develop a remarkable capacity for hallucinatory, vivid, visual experiences known as Eidetic Phenomena. Such phenomena may also involve the senses of touch and hearing and are eliminated through the administration of parathyroid extracts. Thus, normally functioning parathyroids tend to protect us from unsolicited, spontaneous visual and other types of imagery which can so dominate the objective consciousness as to render us incapable of coping with the exigencies of life on this plane.

The excitability of the nervous systems of children demonstrating Eidetic Phenomena has been found to be greatly increased. In one study in Marburg, Germany, during the early 1900's, a number of schoolchildren demonstrating the phenomena were found to come from districts where the water supply was virtually free of calcium. It thus appeared that these children were not receiving enough calcium in their diets.

Treatment with cod-liver oil and calcium or even with calcium alone, caused a disappearance of the eidetic faculty. Returning the children to their calcium-deficient diets caused the faculty gradually to return.

With regard to calcium metabolism, Eidetic Phenomena may arise as a result of parathyroid deficiency, or as a result of thyroid hyperactivity. In either case blood and tissue calcium levels are so lowered as to render the nervous system hyper-excitable. On the other hand, thyroid insufficiency or parathyroid hyperactivity so increase blood and tissue calcium levels as to render the nervous system so insensitive as to result in mental dullness. Thus we see, once again, that in our quest for self-realization the functioning of our glands may profoundly influence the types of experiences we may have. Nevertheless, the mystic knows that the mental poise and balance that accompany properly functioning glands only provide the framework in which the highest of mystical experiences may be sought and realized.

6 The Pituitary

We have seen that each of the endocrine glands thus far considered has a specific role to play in body harmony, and in so doing exerts profound influences on our experience of self. But although each gland in its individual functionings is impressive in the extent of its hidden, and not-so-hidden influences, little has so far been revealed with regard to the control and coordination of glandular functions. We have frequently alluded to the inner or psychic self, and have intimated that it is this aspect of Self which exerts the major regulatory influence on the functionings of the glands. On the other hand, we have also intimated that our outer, objective ways of behaving and thinking also exert profound influences on glandular activities and hence on body harmony. How, then, are these two "opposing" agents of influence to be reconciled?

The Pituitary as Instrument of Harmonization

The student of mysticism strives to bring the outer, objective ways of thinking and acting in line with those "noble" urges arising within the psychic self. From the Rosicrucian point of view, the psychic self is in direct contact with that great sea of infinite intelligence and wisdom—the Cosmic. In the human body the instrument through which the psychic self exerts its most direct influence is the Autonomic Nervous System. This "ancient" portion of the nervous system is often referred to as the Sympathetic Nervous System because it vibrates in "sympathy" with those "higher" vibrations known as Cosmic vibrations. As a reminder of this important harmonic relationship between the psychic self and the Autonomic Nervous System we will continue to use the term "Sympathetic Nervous System."

It is important to realize that because all is of the Cosmic, the Sympathetic Nervous System also vibrates in "sympathy" with the outer self. And since each endocrine gland receives nerve impulses directly from the "Sympathetic Nervous System," it is easy to see how the inner and outer selves each exert an influence on glandular function. Thus, the degree of harmony which exists among our endocrine glands may often be reflective of the degree of harmony which we permit to exist between our inner and outer selves.

But what does all of this have to do with control and coordination of activities among the endocrines? The fact is, while the inner or psychic self is "on the job" twenty-four hours per day, the outer or objective self is usually much less consistent in its efforts. Further, the psychic self has at its disposal the infinite wisdom of the Cosmic while the objective self ordinarily has access only to its limited experience. As a consequence, the vital functions of the body are controlled and regulated by the psychic self which tries to maintain the highest possible degree of harmony within each gland, and among the family of endocrine glands. The pituitary gland, also thought by many to be "the master gland," represents the lower end of the "sympathetic" system through which the psychic self directly controls and coordinates the entire glandular system.

Structure and Location of the Pituitary

Like the adrenal gland, the pituitary is derived from two types of tissue. Let us recall that each adrenal gland consists of an outer portion or cortex and an inner portion or medulla. Let us recall also that the adrenal cortex arises from the same embryonic region as do the gonads, while the medulla consists of cells from the Sympathetic Nervous System. In an analogous manner, the pituitary gland is a double gland arising from two types of embryonic tissue. The front portion of the pituitary gland is known as the *anterior pituitary* and arises from the same embryonic region as does the thyroid gland. On the other hand, the back or hind portion of the pituitary gland is known as the *posterior pituitary* and is actually an outgrowth of the base of the brain. And just as the two portions of each adrenal gland have distinct but complementary functions, so too the anterior and posterior pituitary have distinct and often complementary influences.

In the human, the pituitary gland is located in the center of the head on a line with the base of the bridge of the nose and just above the back portion of the roof of the nasal cavity. See Figure 6-1. The posterior pituitary is connected by a stalk to that portion of the brain known as the *hypothalamus.* The hypothalamus is very closely associated with the pineal gland which serves as the "head" of the Sympathetic Nervous System. This indirect link between the pineal and the

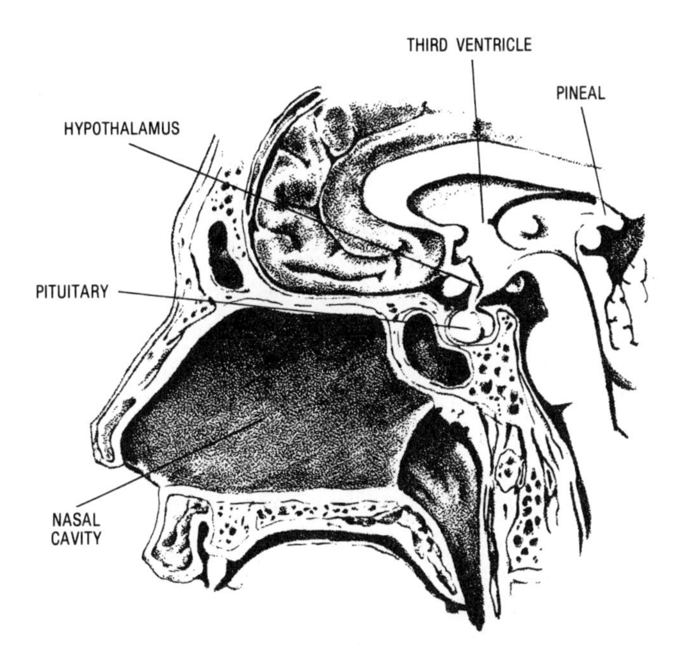

Figure 6-1. Diagram showing the location of the *Pituitary Gland* in relation to the nasal cavity, hypothalamus, *Pineal Gland* and the third ventricle of the brain.

pituitary via the hypothalamus will be dealt with in the next chapter. Suffice it to say at this point that in its coordination of the activities of other endocrine glands, the pituitary is itself controlled by "higher" centers in the brain.

Physiological Functions of the Anterior Pituitary

Perhaps the most obvious effects of pituitary malfunction are to be recognized in cases of dwarfism, gigantism, and acromegaly. The anterior pituitary produces and releases into the blood stream a hormone known as growth hormone (GH). Under ordinary circumstances, sufficient growth hormone gets to all tissues of the body to insure balanced and harmonious growth. When enough of this hormone does not get to the body tissues, especially bone, during the childhood period of life, then growth is minimal and the result is pituitary dwarfism. On the other hand, should too much growth hormone get to the body tissues, then growth is accelerated and the individual becomes a "giant."

As has been discussed in the chapter on the gonads, during adolescence the onset of gonadal hormone activity eventually causes the growing points of bones to fuse. Under ordinary circumstances this fusion brings a halt to further growth in bone tissue since the amounts of growth hormone reaching the tissues is not sufficient to cause any net increase in body height. However, should the anterior pituitary continue to produce large amounts of growth hormone, due to a pituitary tumor for example, then the bones begin to gain in bulk though not in length. This increase in mass becomes most obvious in the hands, feet, and face, and can lead to the deformities of acromegaly. Thus we see that through the activities of growth hormone the anterior pituitary exerts considerable influence on the size of the body and its many organs. However, the anterior pituitary also exerts direct influences on many of the other endocrine glands.

There are a group of anterior-pituitary hormones known as *tropic hormones* because they stimulate the activities of other glands. Individual tropic hormones stimulate the activities of the thyroid gland, the adrenal cortex and the gonads. With regard to the gonads, the timing of release of the *gonado-tropic hormones* by the anterior pituitary is crucial for the onset of puberty.

Thus, a late release of gonado-tropic hormones prolongs the childhood period of life, while an early release shortens this period. The "timing of release" is not determined by the pituitary but by "higher" brain centers. It is interesting to note that the activities of the thymus gland are also intimately associated with the timing of onset of puberty.

On the basis of what has been said previously concerning the influence of thyroid function with regard to mental alertness and mental agility, it is obvious that failure of the anterior pituitary to release the thyroid stimulating *thyrotropic hormone* would result in drastically reduced thyroid activity which, depending upon the period of life, would lead to cretinism or myxedema with their characteristic mental dullness. On the other hand, elevated amounts of thyro-tropic hormone could be expected to lead to thyroid hyperactivity with its consequent nervousness, irritability and over-energetic mental, emotional, and physical responses.

As was indicated in the chapter on the adrenals, subnormal function of the adrenal cortex has profound influences on some aspects of behavior. A case in point relates to the extreme hypersensitivity of subjects suffering from Addison's Disease, a disorder generally characterized by an insufficiency in cortical function of the adrenal. Such adrenal insufficiency often results from inadequate stimulation of the adrenal cortex due to low levels of the cortico-tropic hormone from the anterior pituitary. On the other hand, excess adrenal cortico-tropic hormone produces a hyperactive adrenal cortex with a corresponding lowering of sensitivity.

Simply on the basis of the indirect influence of the tropic hormones on the body in general, and on the nervous system in particular, the anterior pituitary may be regarded as the "gland of intelligence," for it is through pituitary responsiveness, or lack of it, that the intelligence of the psychic self may or may not manifest in our outer, objective behavior. Accordingly, if for genetic or other reasons the anterior pituitary functions subnormally we may have an individual who, in the outer objective world, is labelled a dullard, an idiot, a cretin! Conversely, should the anterior pituitary be hyperactive in its functions, depending upon the degree of hyperactivity, we may have a genius or someone so sensitive to stimuli that outer, objective existence becomes virtually impossible.

The anterior pituitary produces and secretes hormones other than growth hormone and the tropic hormones so far considered. These hormones are *prolactin* and *MSH**. Prolactin (PRL)

*MSH is an abbreviation for *M*elanophore *S*timulating *H*ormone.

initiates and maintains milk production but only after the mammary glands of the breasts have first been sensitized by the higher levels of female sex hormones that characterize pregnancy. Again, the timing of Prolactin (PRL) release by the anterior pituitary is under the control of "higher" centers in the brain.

In lower animals MSH triggers the dispersion of pigment granules in the skin and therefore, causes a darkening of the skin. MSH is probably involved in changes in skin coloration which some animals produce in response to their environment. In the human, MSH brings about a darkening of the skin usually in response to prolonged exposure to sunlight. However, there are rare occasions on which tumors arising among those pituitary cells that produce MSH do cause the overproduction of this hormone. During the mid 1970's a celebrated case involving such a tumor concerned a white South African lady who, within a few months, underwent such a complete color change, that on the basis of skin color alone she was indistinguishable from the average black person. Nevertheless, apart from such dramatic influences on skin coloration, MSH enables mammals, including humans, to be more attentive, particularly to visual stimuli. In addition, MSH has been found to improve learning in mentally retarded individuals.

Physiological Functions of the Posterior Pituitary

Apart from differences with regard to embryonic origins, the posterior pituitary differs from its companion in other important ways. Thus, unlike the hormones of the anterior pituitary, the two known hormones of the posterior pituitary are not locally manufactured. Instead, posterior pituitary hormones are manufactured by the cells of the hypothalamus and transported to the posterior pituitary where they are stored.

The hormones of the posterior pituitary both act on smooth muscles. The two differ slightly in regard to their chemical structure as well as in regard to the type of smooth muscle upon which they act. One of these hormones, *oxytocin* by name, acts on the smooth muscle of the uterus (womb) and the cells of the mammary glands. In acting on the womb, oxytocin is an important factor in the uterine contractions necessary in childbirth. On the other hand, in acting on the mammary glands, oxytocin causes the release of milk necessary for nourishing the newborn. In this regard, the anterior and posterior pituitary function in harmony in that one oversees the development of the infant body and the milk to sustain it, while the other oversees the transition called birth, plus the release of nourishment for the newborn.

The second posterior pituitary hormone is known as *vasopressin* and causes contraction of the smooth muscle of arteries. In lower animals, vasopressin induces an increase in blood pressure by causing the arteries to constrict. In the human, however, vasopressin exerts its most significant influence in regulating the rate at which water is reabsorbed into the body by the kidneys. In the absence of vasopressin, urine output may increase at least tenfold. This condition of increased urine output is known as *diabetes insipidus.* Too great a loss in body water can lead to dehydration which we experience as thirst. The experience of thirst is controlled by the hypothalamus.

Pituitary Not the Master Gland

Overall, the activities of the pituitary gland influence every aspect of life's expressions in the human body. This pervasive influence of the pituitary is graphically represented in Figure 6-2. But despite its obvious importance as a channel through which the psychic self maintains the best possible harmony within the body, the pituitary is not *the* master gland. As indicated earlier, such important aspects as the timing of onset of puberty, or the time of birth of a baby, or even the maintenance of body temperature, to mention but a few, are all activities which fall

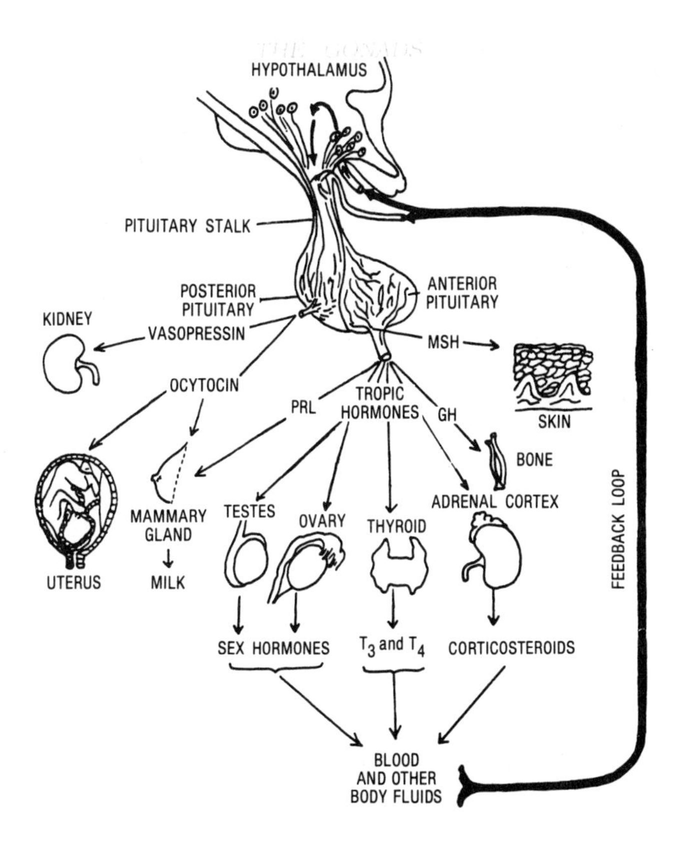

Figure 6-2. Diagrammatic representation summarizing the endocrine activities of the pituitary gland. Hormones of the anterior pituitary are known as *tropic hormones.* Tropic hormones act directly to stimulate the hormonal activities of other glands whose hormone levels in the blood act on the neuro-humor-releasing cells of the hypothalamus. In this way, a feedback loop is established.

beyond the control of the pituitary gland. In order to answer questions such as "how does the pituitary know when to release a given hormone?" or even "how does it know that it has released enough, too little, or too much of a particular hormone?" let us now turn to the activities of the hypothalamus.

7 The Hypothalamus

Although in classical treatments of the endocrine system the hypothalamus is not regarded as an endocrine gland, more recent discoveries indicate that it eminently qualifies to be so considered. As intimated earlier, the hypothalamus manufactures and secretes hormones. At the present time it is not clear if the hypothalamus secretes any of its hormones directly into the general circulation. However, it is known that this organ secretes many of its hormones into the special blood system of the pituitary gland. It is also known that many hypothalamic hormones are regulators of pituitary activity. These special hormones have been referred to as *neuro-humors.*

The hypothalamus is part of the floor of the third chamber of the brain. See Figure 6-1.

Being that portion of the brain proper which is in closest contact with the pituitary gland, the hypothalamus serves as a bridge linking the activities of the Central Nervous System with those of the glandular system. In addition, being closely associated with the "Sympathetic Nervous System" the hypothalamus also serves as a bridge linking the activities of the psychic self with those of the glandular system. Thus, the hypothalamus responds to a blend of impulses representing the best possible balance between the urges from the inner and outer selves. In so responding, the hypothalamus releases "neuro-humors" which direct the activities of the pituitary gland.

Hypothalamic Control of Pituitary Function

Hormones of the hypothalamus are of two types, namely *releasing factors* and *inhibiting factors.* Releasing factors induce the anterior pituitary to release specific hormones, while inhibiting factors prevent the release of hormones from the anterior pituitary. Thus, there are releasing factors which direct the release of the tropic hormones which act on the thyroid, the adrenal cortex, and the gonads. In addition there are release and inhibiting factors for growth hormone (GH) and MSH, as well as for the "milk hormone" prolactin (PRL).

The growth hormone inhibiting factor merits special mention. This hormone called *somatostatin* also inhibits the release of the thyroid-stimulating hormone and acts directly upon the pancreas to suppress the release of the hormones insulin and glucagon. In addition, several aspects of the digestive process are directly influenced by the action of somatostatin. For example, somatostatin is known to inhibit the release of enzymes that are intimately associated with the digestion of food. Thus, through the action of somatostatin, the hypothalamus not only controls the digestion of food but, no doubt, also governs our experience of hunger.

With respect to control of the activities of the posterior pituitary, impulses arising within the hypothalamus are transmitted directly by nerve fibers to the posterior pituitary. These impulses signal the release of stored hormones.

Some impulses arising within the hypothalamus are initiated by the levels of certain chemicals in the blood reaching the hypothalamus. For example, if the levels of certain mineral salts in the blood reaching the hypothalamus are high, then these would be sufficient to cause the hypothalamus to send a message to the posterior pituitary, directing it to release the hormone vasopressin. In its turn, vasopressin causes the kidneys to reduce the amount of water excreted as urine. Such an action by the kidneys has the effect of retaining water in the blood, thereby reducing the level of salt. At the same time, the individual may experience thirst which induces the drinking of water which results in an almost immediate lowering of the level of salt in the blood.

A similar "feedback" mechanism operates in the release of anterior pituitary hormones. Thus, for example, low levels of thyroid hormones in the blood reaching the hypothalamus induce the release of the appropriate releasing hormone from the hypothalamus. In its turn, this hormone causes the release of anterior pituitary thyroid-stimulating hormone which then stimulates the thyroid to produce and release more of its hormones. In this manner the hypothalamus constantly controls the activities of the glandular system because it is constantly "in touch" with the state of the body through signals reaching it via the blood. Refer again to Figure 6-2.

In addition to release and inhibiting factors, the hypothalamus produces a class of morphine-like chemicals known as *endorphins.* There are good indications that endorphins serve as neurotransmitters or chemical messengers in the Central Nervous System. In fact, while endorphins suppress the excitability of a variety of cells in the Central Nervous System, they also stimulate the activity of a class of neurons known as *pyramidal cells.* Pyramidal cells are located in that portion of the brain called the *limbic system,* and may be associated with experiences of exaltation, ecstasy, and "higher states of consciousness." Endorphins also directly affect our experience of pain.

The Hypothalamus as Bridge

From the foregoing it seems clear that the hypothalamus, as a sort of "bridge" between body and psyche, plays a profound role both in body harmony and in our experience of self. In so doing, the hypothalamus seems to protect us from pain and, at the same time serves as the instrument of a prodigious intelligence which, when we allow ourselves, we may experience as ecstasy, exaltation, and oneness. But despite its obvious importance in the control and coordination of the endocrine system, the hypothalamus is but a link between the psychic or Inner Self and the outer self. What then controls the activities of the hypothalamus? Let us now consider the pineal gland.

8 The Pineal

At least as long as 2000 years ago, the pineal gland was singled out by mystic-scientists as the "seat of the Soul." Some of these early scientists even went as far as to assert that the degree of one's psychic development is proportional to the degree of development of the pineal gland. Even in more modern times this assertion is favored by many students of mysticism. A case in point is represented by the statement that "the pineal is concerned with the transition from the self-centered mentality of childhood to one of maturity and benevolence." Such statements have encouraged many well-meaning aspirants to entertain the notion that the larger the pineal gland, the more psychically evolved the individual. Are there any facts in support of the asserted relationship between the pineal and psychic development?

Rene Descartes, philosopher, mystic, and founder of modern mathematics, referred to the pineal gland as the seat of the rational soul. As a philosopher and mathematician, Descartes is unlikely to have used the term "rational" in a loose or nonspecific manner, since "rational" derives from "ratio" which means to compare. Descartes is reported to have referred to the pineal as "la glande cannairienne" or the gland of knowing. And is "knowing" not based upon comparison?

Traditionally then, the pineal gland has been regarded as an organ of perception, "translating perceived information into humors which pass down tubes to influence the workings of the body," according to Descartes. From this traditional point of view, the pineal represents a portal through which the psychic self exerts very definite influences on the physical self.

From the modern scientific point of view, the pineal is often referred to as "a regulator of regulators," governing many activities of the hypothalamus and the pituitary. This view is based largely on results from animal studies and is not incompatible with what is known of the physiological activities of this gland in the human.

The Pineal as Third Eye

In some lower animals, such as lizards and frogs, the pineal actually consists of two distinct organs, one occupying its usual location in the roof of the third chamber of the brain (refer again to Figure 6-1), the other occupying a position exterior to the brain proper and exhibiting characteristics of an eye. This second or exterior type of pineal organ is generally referred to as a para-pineal organ or *third eye*. Physiological evidence now indicates that in animals lacking an external pineal organ, some areas of the roof of the third chamber of the brain, and perhaps even the regular pineal itself, may be capable of translating light vibrations into behavioral and other types of responses in certain animals.

In higher animals, including the human, the pineal is mainly composed of cells which are themselves not responsive to light. However, in such animals, light received through the eyes seems to influence pineal function since activities known to be associated with the pineal gland are altered in blind animals and are greatly influenced by light in sighted animals. In addition, in higher animals, pineal function is also influenced by impulses from the "Sympathetic Nervous System."

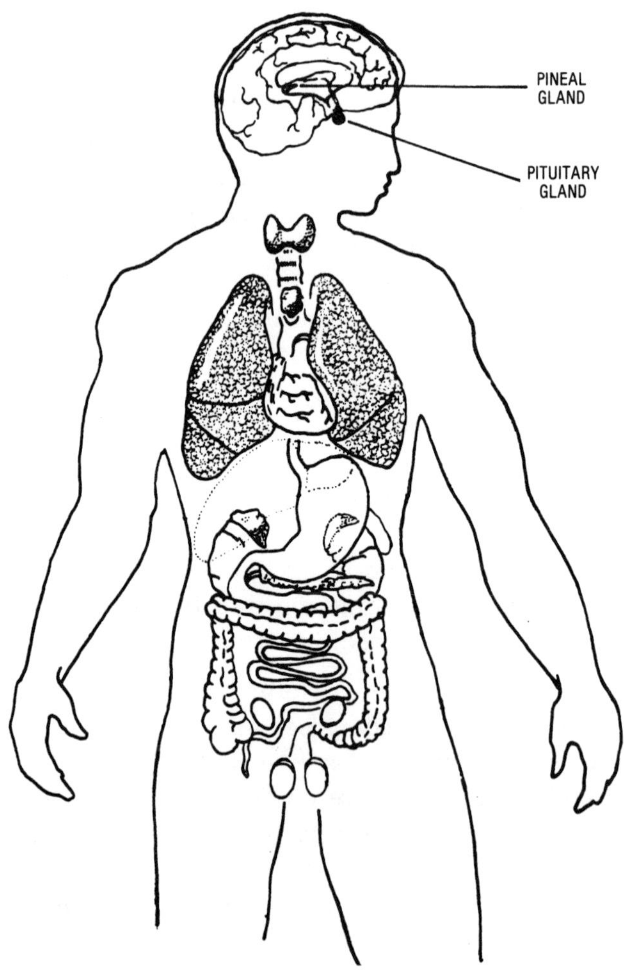

Figure 8-1. Diagram showing the location of the *Pineal Gland* in relation to the *Pituitary Gland* in the human body.

The Pineal as Regulator of Regulators

In the human, the pineal is a small pinecone-shaped organ located at the very center of the head. See Figure 8-1. The pineal remains biochemically active throughout the life of the individual. Among its variety of influences, the pineal affects the activities of the adrenal cortex, the thyroid, the thymus, the gonads, the hypothalamus-pituitary system, and apparently the pancreas. The pineal, therefore, qualifies as the "regulator of regulators."

In the absence of the pineal, several alterations in behavior and brain electrical activity have been observed in many animal species. Thus, the absence of the pineal in young rats leads to an early onset of puberty. In this regard it is interesting to note that in the human, tumors which surround the pineal and, therefore, inhibit its activities, also lead to an early onset of puberty. In such cases puberty occurring at five years of age, for example, is not unusual. On the other hand, hyperactivity of the pineal, caused by tumors within the gland itself, leads to considerable delay in the onset of puberty.

The Pineal and the Transition called Puberty

Because of the particular importance of the transition of puberty for subsequent experiences in the realization of self, it is fitting that we examine a little more closely the influence of the pineal on physical and mental maturity. In terms of the more modern concepts with regard to pineal activity, this gland came into the limelight as a result of the following case history and others like it.

During the early 1900's a five-year old Austrian boy suddenly began to grow quite rapidly, developing in stature and looks to resemble a boy of twelve or thirteen. Hair developed over his body, his voice became low-pitched, and he exhibited all the signs of having gone through puberty. His sexual precociousness was accompanied by a mental precociousness. He is reported to have questioned his parents concerning the fate and condition of the soul after death. On one occasion he is reported to have remarked reflectively: "It is odd how much better I feel when I let other children play with my toys than when I play with them myself." Other statements attributed to this premature young man reflect a maturity of thought and mental process that is normally considered to be the result of many years of adult experience. The child died before the age of six, some four weeks after being admitted to the hospital. Autopsy showed that he had a tumor of the pineal gland.

As indicated earlier, an early onset of puberty, when associated with the pineal, is due to pineal inactivity or hypoactivity, since during the childhood period the activities of the pineal are associated with inhibiting the release of the gonad-stimulating hormones from the anterior pituitary. In the case history reported above, the "tumor of the pineal" must have been of the type which inhibits pineal activity and, therefore, hastened sexual and mental maturity. It seems clear, then, that the timing of release of gonad-stimulating hormones, and, as a consequence, the timing of onset of puberty, is under the control of the pineal. One would suspect therefore, that following puberty there should be some alteration in the activities of the pineal gland. To date, no such alteration in activity, at least from a biochemical point of view, has been demonstrated.

The Pineal and the Intuitive Faculty

It has been observed, however, that following puberty the degree of calcification, i.e., the amount of sandy material present in the pineal, progresses noticeably with age in our technologically dominated society. By contrast, there is considerably less calcification of the pineal in individuals living in cultures which function more in harmony with the cycles of nature. It is also well known that members of a technological culture, on the average, rely less on the intuitive faculty than those of a more "primitive" culture.

There is no direct evidence linking the disuse of the intuitive faculty to the degree of calcification of the pineal, however. Nevertheless, it is tempting to speculate that the degree of calcification of the pineal gland reflects the degree of our alienation from nature and her rhythms.

However, it does appear that some degree of pineal calcification is inevitable. Further, it appears that such changes in pineal structure are necessary if we are to undergo the transition from childhood to adulthood. Thus, in becoming adults it may be necessary for us to sacrifice some degree of the intuitive faculty in order to acquire a degree of reason—a degree of the rational. Nonetheless, how much of the intuitive faculty we eventually sacrifice in order to depend mainly on the faculty of reason seems to be a matter of choice. Hence, not *all* adult members of our technological society have highly calcified pineal glands. And yet, from a biochemical point of view, the activities of the individual pineal gland do not change with age. This clearly indicates that, from the point of view of psychic abilities, the most sophisticated chemical methods available at this time are not capable of indicating just where one is on the "scale" of psychic development.

Pineal Hormones

The variety of influences that the pineal gland exerts on the entire glandular system, and hence on the physical sense of well-being which the individual experiences, indicates that this "gland of glands" must make its influences felt via some hormone or hormones. In this regard there are two known pineal hormones whose activities pretty well account for the known influences of the pineal gland on the physical body. Accordingly, the hormone *melatonin* has been shown to produce many of the influences known to be mediated by the hypothalamus. These activities include the release of the tropic or stimulating hormones from the anterior pituitary.

On the other hand, the pineal hormone *vasotocin* delays childbirth, indicating that it interferes with the release of the posterior pituitary hormone, oxytocin. It is interesting, therefore, to note that the pineal has been implicated in synchronizing pregnancy and the birth of the young with the season (spring) which is maximally conducive to the raising and survival of young animals. Thus, it appears that modern findings tend to support the mystical view that the pineal may indeed function as the doorway through which psychic influences are made to manifest in the body.

The Pineal and Light-Dark Rhythms

As indicated earlier, pineal activities, directly or indirectly, are responsive to environmental lighting. In many birds and mammals, light appears to be a synchronizer of animal physiology and behavior. Thus, many physiological cycles and behavioral cycles are oriented around the daily alternation of light and darkness. Such daily cycles are called *circadian rhythms.*

In the human, and other mammals, circadian rhythms manifest as fluctuations in blood levels of the adrenal cortex hormones that are associated with periods of activity. Of even greater significance is the fact that in subjects blind because of cataracts, the degree of cyclic change in such hormones is greatly reduced. Following successful cataract removal, however, the amplitude of the adrenal hormone rhythms returns to normal.

It thus appears that the actual visual experience of light has profound effects on our glandular system. In this regard, it is of interest to note that rats maintained in constant light have *smaller* pineal glands than comparable ones kept in constant darkness. By contrast, hens kept in constant light had *larger* pineals than those kept in constant darkness. But since rats are nocturnal, i.e., their periods of most intense activity occur during darkness while chickens are most active during periods of light, the relationship between relative pineal size and environmental lighting during periods of maximal activity is apparent. It seems clear, then, that the pineal gland is an important organ linking the environment with, and through, the endocrine system via the eyes and other organs of perception. Thus, the pineal may be the link between the macrocosm or larger universe, on the one hand, and man, the microcosm or smaller universe on the other. The pineal may, therefore, indeed be the "eye" through which man harmonizes both the inner and outer worlds.

Having now considered each gland of the endocrine system, what have we learnt that may lead us to a clearer understanding of the psychic self? We have had much to say about the physiological functions of the glands, and have clearly demonstrated how these physiological functions affect our mentality. But what is the distinction between "mental" and "psychic"? How does the functioning of our glands affect our psychic nature? In order to arrive at a clearer understanding of the relationship between the Inner and outer selves, let us briefly consider the role of the glands in psychic life.

9 The Mirror of Self

Although we have treated each gland as though it were a thing apart, it is important to bear in mind that in actuality the glands function together as an integrated whole. Neglect in considering the integral wholeness of the system of glands leaves us open to the notion that one or another of the glands is more important than the others. That no one gland is more important than another in the process of self-realization is clearly indicated by the fact that each has its role to play. In addition, any influence exerted on any one of the glands soon makes itself felt throughout the entire system. As may be expected, the system of glands may be influenced by phenomena occurring within or without the individual.

From a Rosicrucian point of view, by far the most important influence that can be exerted upon the glandular system is associated with activities occurring within the psychic self. For this reason we refer to the system of glands as the "Mirror of Self." But what exactly do we mean by the term Self?

Ordinarily, when we speak of Self we refer to a realization of our inner nature. In other words, we do not speak of Self in its actuality. Rather, we speak of a reflection, and from this reflection we speculate as to the true nature of Self. In our speculations we often come to conclude that the quality of Self's reflection must be determined by the quality of the mirror. And thus, because of our understanding of the law of cause and effect, we surmise that if we could improve the quality of the mirror we would succeed in improving the quality of the reflection as well. However, this is only partly true.

In attempting to account for many of the obvious inequities of human existence, we are often seduced by facile explanations which hold, for example, that there is a direct causal relationship between the quality of the "mirror" on the one hand, and the quality of the Self being reflected, on the other. Although such explanations are logical from the limited point of view of objective intellect, we must still face the thorny problem raised by those outstanding human beings who, although sophisticated chemical and physiological analyses show their "mirror" to be just average, shine forth as beacons of courage and virtue in the annals of history. It appears then, that in addition to the physical qualities of the mirror, there must be some other factor which determines the quality of the Self's reflection. This other factor is the brilliance of the Light which we allow to fall upon the mirror. Whence cometh this Light? And how do we influence the quantity that will fall upon the mirror?

Following the nine months of gestation during which the infant body has been duly prepared under the guidance of the intelligence inherent in the mother's being, a child is born. From a Rosicrucian point of view this phenomenon of birth is characterized by the entry of a soul personality into the infant's body. A soul personality has within it the Divine Intelligence of the Cosmic, plus complete memory insofar as all human and personal experiences are concerned. As the child grows and the personality unfolds in accordance with the limitations imposed by the physical and physiological characteristics of the body, patterns of behavior begin to emerge. These patterns of behavior are reflections of past experience—past habits developed by the individual in response to particular situations. These patterns of behavior are regenerated and perpetuated until such time as the individual decides to change them. Thus, if the individual has developed the habit of turning his back on the wisdom, the Light residing at the very core of his personality, then this habit will be passed on from one incarnation to another just as a mutant gene is regenerated and perpetuated from one generation to another via the sex cells. In a similar manner "good" habits are carried within the personality from one incarnation to another. We see, therefore, that the principle of regeneration manifests within the psychic self of the soul personality just as it does in the physical self.

Developed patterns of behavior constitute character. In turn, character influences the flow of Spirit Energy within the psychic self. When behavior patterns are in harmony with the Cosmic flow at the core of the soul personality, then the intensity of the Light flowing through the psychic self is great. And since the psychic self uses the glandular system as its "mirror," the degree of harmony existing within the character is reflected in the body. How then do we cultivate harmony within the psychic self?

As Rosicrucians we begin with the heart—the God of our Heart. Just as the physical heart represents the force behind the distribution of nourishment throughout the physical self, so too the God of our Heart represents the force behind the distribution of spiritual nourishment throughout the psychic self. And just as the thymus gland is the "heart center" through which the integrity of the body's sense of self is maintained, so too the psychic "heart center" maintains the integrity of the psychic sense of self. As a symbol of our highest ideals and aspirations, the God of our Heart draws Light into the psychic self just as the action of the physical heart draws certain aspects of the vitalizing force of NOUS into the physical self. Hence, the intensity of Light flowing through the psychic self is a reflection of the quality and intensity of the idealistic aspirations we hold within our heart of hearts.

The intensity of desire soon finds expression in that outward movement of psychic energy which we call emotion. This outward movement is reflected in the activities of the adrenal glands which, in the physical body, are associated with the mobilization of physical energy. Thus, just as the hormones of the adrenal glands mobilize the cells of the thymus gland in the defense of the physical self, so too desire mobilizes the emotions in defense of the psychic self. Because of this correspondence between the emotions and desire it becomes possible through analysis to determine the nature of our ideals. Accordingly, the higher emotions may be said to reflect the true nature of the "God of our Heart," while the "baser" emotions may be said to reflect the nature of the "false gods" or pretenders.

In contemporary scientific circles the adrenal glands are often referred to as the "glands of fight or flight." The implication here is that fight and flight are the major, if not the exclusive, uses to which the enormous amounts of energy, made available to the individual by the activities of the adrenal glands, can be put. A moment's thought soon reveals that such an implication can only be valid if the ideal of limited self-preservation is at the heart of our every thought and action. The fact is that the activities of the adrenals are also associated with love, and ecstasy and even with the afflatus of the soul which Rosicrucians call Peace Profound. Thus, the nature of an emotional experience is related to the nature of the ideal to which we are paying homage at the time of the experience.

Although the nature of the ideal may be the same for a number of individuals, the intensity of the corresponding emotional experience may vary considerably from one individual to another. Even within a given individual the intensity of a particular emotional experience may vary from one occasion to another even though the intensity of the desire is comparable. Why should this be so?

At the outset it must be realized that the intensity of an emotion is related to the rate at which psychic energy is allowed to manifest in the outer consciousness. Thus, a rapid release leads to a short-lived but intense realization while a slow release leads to a sustained but mild realization. In the physical body the activities of the thyroid gland are associated with the rate at which energy is released into the system. In an analogous manner, the activities of a corresponding psychic function controls the rate at which psychic energy moves outward into objective consciousness.

The level of thyroid activity, in its physical and psychic manifestations, is not the same among all individuals. Thus, just as the rate of energy release within the body is accelerated in individuals with hyperactive thyroids, an analogous situation can exist within the psychic self. Conversely, a psychic sluggishness can exist in those individuals whose psychic thyroid function is subnormal. Rosicrucian exercises are designed to harmonize both psychic and physical functions of the glandular system.

The ideals we choose to govern our lives have profound influences on the type of ideas and concepts and even experiences, which we will choose during the course of our spiritual unfoldment. Just as with physical nourishment these ideas, concepts, and experiences are broken down into their simplest components which are then stored or assembled into larger components in accordance with the needs of the psychic self. Thus, the formation of habits finds reflection in the dual process of digestion-assimilation which is largely dominated by the activities of the pancreas. And just as the types of food we eat determine the variety and subtlety of the physical energies of the body, so too the habits we form on the basis of chosen experiences determine the variety and subtlety of the energies flowing through the psychic self. Bad eating habits often lead to physical and psychic ill-health.

But energy mobilization, availability, and rate of release does not necessarily speak of intelligent utilization. If an integrated system is to be self-sustaining then it must be self-monitoring and self-regulating. The self-monitoring aspect finds expression in the activities of the hypothalamus-pituitary axis of the brain. As will be recalled, this portion of the brain is exquisitely sensitive to the influences of the other glands. Because of this sensitivity the hypothalamus and pituitary release appropriate regulating hormones in response to the level of activity of each gland. However, lest we conclude that the hypothalamus-pituitary is more important than the other glands, let us remember that in harmonizing and regulating the activities of the other glands, the activities of the pituitary and hypothalamus are themselves regulated and harmonized. Such a system of reciprocal regulation is a reflection of the infinite wisdom of the Cosmic. A similar harmonizing intelligence manifests in the psychic self and embodies the principle of self-consciousness.

It is interesting to note, however, that of all the glands, no hypothalamus-pituitary regulation of the thymus has been established. This relative lack of direct regulatory influence assumes particular importance when it is recalled that some hormones of the thymus exert profound influences on the activities of the pituitary gland. In fact, as the heart of the body's sense of self, the thymus seems to have more of an affinity with the pineal gland than with the pituitary-hypo thalamus. This may well be a reflection of the complementary relationship which exists between the "wisdom of the head" and the "wisdom of the heart."

All in existence tends toward self-preservation. Nevertheless, all in existence is also forever changing, forever becoming something else. Between these two tendencies the drama of life unfolds. Together these two tendencies represent the head and the tail, the beginning and the end of the circle—the totality of BEING.

Now, as has already been indicated, the tendency toward self-preservation or self-perpetuation manifests at the level of the gonads. However, little has yet been said concerning the tendency toward becoming something else.

In its endocrine function the pineal gland serves as a synchronizer which synchronizes the cycles of birth as well as the cycles of physical and mental unfoldment. In addition, the activities of the pineal are associated with synchronizing the rhythmic activity of each gland as well as with the integrated rhythm of the entire glandular system. Further, through its sensitivity to light, the pineal synchronizes the rhythm of the glandular system with the rhythms and cycles of the entire cosmos. In an analogous manner, through its sensitivity to psychic light, the psychic pineal links and synchronizes the rhythms of the psychic self with those sublime rhythms that characterize the law and order of the Cosmic.

However, as may be expected, the degree of harmony existing between the rhythmic action of the psychic self and the Cosmic is determined by the quality of the ideals we choose to serve. Thus, if at the heart of our sense of self we hold thoughts of wholeness, oneness, and unity with all that exists, or could exist, because the Cosmic is the ALL, then we are in harmony with the Cosmic, which favors none over another. And since it is only through harmony and resonance that energy is transmitted from one state to another, then when we resonate in harmony with the Cosmic, we become divinely conscious.

The task of the student is, therefore, to discover a personal Path to the goal of harmonious thinking, acting, and living. This Path is dual in nature, possessing both inner and outer attributes. The outer attributes include an adequate mirror which, through its reflective qualities, often serves as a trustworthy indicator to the student on the Path as to the success or failure of directed effort. Nevertheless, we should not lose sight of the fact that often the spiritual aspect of Self chooses certain "defective mirrors" in order that the psychic self may experience certain well-defined aspects of its own awareness. Such choices defy the lure of facile explanations and leave us wondering, "Why was I born into these circumstances? Would it not have been better had I been born rich and ugly rather than poor and handsome?" The sensible answer to such questions can only be: "Perhaps, but why not discover how you may best utilize the circumstances of your birth to attain even greater harmony with the God of your Heart? Is not the admonition 'if you have lemons, make lemonade' sage advice?"

Explanatory
THE ROSICRUCIAN ORDER
Purpose and Work of the Order

Anticipating questions which may be asked by the readers of this book, the publishers take this opportunity to explain the purpose of this Order and how you may learn more about it.

There is only one universal Rosicrucian Order existing in the world today, united in its various jurisdictions, and having one Supreme Council in accordance with the original plan of the ancient Rosicrucian manifestoes. The Rosicrucian Order is not a religious or sectarian society.

This international organization retains the ancient traditions, teachings, principles, and practical helpfulness of the Order as founded centuries ago. It is known as the *Ancient Mystical Order Rosae Crucis,* which name, for popular use, is abbreviated into AMORC. The Headquarters of the Worldwide Jurisdiction (The Americas, Australasia, Europe, Africa, and Asia) is located at San Jose, California.

The Order is primarily a humanitarian movement, making for greater Health, Happiness, and Peace in people's *earthly lives*, for we are not concerned with any doctrine devoted to the interests of individuals living in an unknown, future state. The Work of Rosicrucians is to be done *here* and *now*; not that we have neither hope nor expectation of *another* life after this, but we *know* that the happiness of the future depends upon *what we do today for others* as well as for ourselves.

Secondly, our purposes are to enable men and women to live clean, normal, natural lives, as Nature intended, enjoying *all* the privileges of Nature, and all benefits and gifts equally with all of humanity; and to be *free* from the shackles of superstition, the limits of ignorance, and the sufferings of avoidable *Karma*.

The Work of the Order, using the word "work" in an official sense, consists of teaching, studying, and testing such Laws of God and Nature as make our Members Masters in the Holy Temple (the physical body), and Workers in the Divine Laboratory (Nature's domains). This is to enable our Members to render *more efficient help* to those who do not know, and who need or require help and assistance.

Therefore, the Order is a School, a College, a Fraternity, with a laboratory. The Members are students and workers. The graduates are unselfish servants of God to Humanity, efficiently educated, trained, and experienced, attuned with the mighty forces of the Cosmic or Divine Mind, and Masters of matter, space, and time. This makes them essentially Mystics, Adepts, and Magi—creators of their own destiny.

There are no other benefits or rights. All Members are pledged to give unselfish Service, without other hope or expectation of remuneration than to Evolve the Self and prepare for a *greater* Work.

The Rosicrucian Sanctum membership program offers a means of personal home study. Instructions are sent once a month in specially prepared weekly lectures and lessons, and contain a summary of the Rosicrucian principles with such a wealth of personal experiments, exercises, and tests as will make each Member highly proficient in the attainment of certain degrees of mastership. The lectures are under the direction of the Imperator's staff. These correspondence lessons and lectures comprise several Degrees. Each Degree has its own Initiation ritual, to be performed by the Member at home in his or her private home sanctum. Such rituals are not the elaborate rituals used in the Lodge Temples, but are simple and of practical benefit to the student.

If you are interested in knowing more of the history and present-day helpful offerings of the Rosicrucians, you may receive a *free* copy of the booklet entitled *Mastery of Life*, by sending a request to:

Scribe G.K.A.
Rosicrucian Order, AMORC
Rosicrucian Park
San Jose, California 95191, U.S.A.

ROSICRUCIAN LIBRARY

ROSICRUCIAN QUESTIONS AND ANSWERS WITH COMPLETE HISTORY OF THE ORDER

by H. Spencer Lewis, F.R.C., Ph. D.

From ancient times to the present day, the history of the Rosicrucian Order is traced from its earliest traditional beginnings. Its historical facts are illuminated by stories of romance and mystery.

Hundreds of questions in this well-indexed volume are answered, dealing with the work, benefits, and purposes of the Order.

ROSICRUCIAN PRINCIPLES FOR THE HOME AND BUSINESS

by H. Spencer Lewis, F.R.C., Ph. D.

This volume contains the practical application of Rosicrucian teachings to such problems as: ill health, common ailments, how to increase one's income or promote business propositions. It shows not only what to do, but what to avoid, in using metaphysical and mystical principles in starting and bringing into realization new plans and ideas.

Both business organizations and business authorities have endorsed this book.

THE MYSTICAL LIFE OF JESUS

by H. Spencer Lewis, F.R.C., Ph. D.

A full account of Jesus' life, containing the story of his activities in the periods not mentioned in the Gospel accounts, *reveals the real Jesus* at last.

This book required a visit to Palestine and Egypt to secure verification of the strange facts found in Rosicrucian records. Its revelations, predating the discovery of the Dead Sea Scrolls, show aspects of the Essenes unavailable elsewhere.

This volume contains many mystical symbols (fully explained), photographs, and an unusual portrait of Jesus.

THE SECRET DOCTRINES OF JESUS

by H. Spencer Lewis, F.R.C., Ph. D.

Even though the sacred writings of the Bible have had their contents scrutinized, judged, and segments removed by twenty ecclesiastical councils since the year 328 A.D., there still remain buried in unexplained passages and parables the Great Master's *personal* doctrines.

Every thinking man and woman will find *hidden truths* in this book.

"UNTO THEE I GRANT..."

as revised by Sri Ramatherio

Out of the mysteries of the past comes this antique book that was written two thousand years ago, but was hidden in manuscript form from the eyes of the world and given only to the Initiates of the temples in Tibet to study privately.

It can be compared only with the writings attributed to Solomon in the Bible of today. It deals with man's passions, weaknesses, fortitudes, and hopes. Included is the story of the expedition into Tibet that secured the manuscript and the Grand Lama's permission to translate it.

A THOUSAND YEARS OF YESTERDAYS

by H. Spencer Lewis, F.R.C., Ph.D.

This fascinating story dramatically presents the real facts of reincarnation. It explains how the soul leaves the body and *when* and *why* it returns to Earth again.

This revelation of the *mystic laws and principles* of the Masters of the East has never before been presented in such a form. Finely bound and stamped in gold, it makes a fine addition to your library.

SELF MASTERY AND FATE WITH THE CYCLES OF LIFE

by H. Spencer Lewis, F.R.C., Ph. D.

This book demonstrates how to harmonize the self with the cyclic forces of each life.

Happiness, health, and prosperity are available for those who know the periods in their own life that enhance the success of varying activities. Eliminate "chance" and "luck," cast aside "fate," and replace these with self mastery. Complete with diagrams and lists of cycles.

ROSICRUCIAN MANUAL

by H. Spencer Lewis, F.R.C., Ph.D.

This practical book contains useful information that complements your Rosicrucian monograph studies. Included are extracts from the Constitution of the Rosicrucian Order, an outline and explanation of Rosicrucian customs, habits, and terminology, diagrams that illustrate important mystical principles, explanations of symbols used in the teachings, biographical sketches of AMORC officials, a glossary, and other helpful material. This book will answer many questions you may have about AMORC and its teachings, whether you are a Neophyte student or a member studying in the higher Degrees. The *Rosicrucian Manual* has been expanded and updated since its first printing to better serve our members through the years.

MYSTICS AT PRAYER

Compiled by Many Cihlar, F.R.C.

The first compilation of the famous prayers of the renowned mystics and adepts of all ages.

The book *Mystics at Prayer* explains in simple language the reason for prayer, how to pray, and the Cosmic laws involved. You come to learn the real efficacy of prayer and its full beauty dawns upon you. Whatever your religious beliefs, this book makes your prayers the application not of words, but of helpful, divine principles. You will learn the infinite power of prayer. Prayer is man's rightful heritage. It is the direct means of man's communion with the infinite force of divinity.

BEHOLD THE SIGN

by Ralph M. Lewis, F.R.C.

Unwrap the veil of mystery from the strange symbols inherited from antiquity. What were the *Sacred Traditions* said to be revealed to Moses? What were the discoveries of the Egyptian priesthood?

This book is fully illustrated with *age-old secret symbols* whose true meanings are often misunderstood. Even the mystical beginnings of the *secret signs* of many fraternal brotherhoods today are explained.

MANSIONS OF THE SOUL

by H. Spencer Lewis, F.R.C., Ph. D.

Reincarnation—the world's most disputed doctrine! What did Jesus mean when he referred to the "mansions in my Father's house"? This book demonstrates what Jesus and his immediate followers knew about the rebirth of the soul, as well as what has been taught by sacred works and scholarly authorities in all parts of the world.

Learn about the cycles of the soul's reincarnations and how you can become acquainted with your present self and your past lives.

LEMURIA—THE LOST CONTINENT OF THE PACIFIC

by Wishar S. Cervé

Where the Pacific now rolls in a majestic sweep for two thousand miles, there was once a vast continent known as Lemuria.

The scientific evidences of this lost race and its astounding civilization with the story of the descendants of the survivors present a cyclical viewpoint of rise and fall in the progress of civilization.

THE TECHNIQUE OF THE MASTER
or The Way of Cosmic Preparation

by Raymund Andrea, F.R.C.

A guide to inner unfoldment! The newest and simplest explanation for attaining the state of Cosmic Consciousness. To those who have felt the throb of a vital power within, and whose inner vision has at times glimpsed infinite peace and happiness, this book is offered. It converts the intangible whispers of self into forceful actions that bring real joys and accomplishments in life. It is a masterful work on psychic unfoldment.

THE SYMBOLIC PROPHECY OF THE GREAT PYRAMID

by H. Spencer Lewis, F.R.C., Ph. D.

The world's greatest mystery and first wonder is the Great Pyramid. Its history, vast wisdom, and prophecies are all revealed in this beautifully bound and illustrated book. You will be amazed at the pyramid's scientific construction and at the secret knowledge of its mysterious builders.

THE TECHNIQUE OF THE DISCIPLE

by Raymund Andrea, F.R.C.

The Technique of the Disciple is a book containing a modern description of the ancient, esoteric path to spiritual Illumination, trod by the masters and avatars of yore. It has long been said that Christ left, as a great heritage to members of His secret council, a private method for guidance in life, which method has been preserved until today in the secret, occult, mystery schools.

Raymund Andrea, the author, reveals the method for attaining a greater life taught in these mystery schools, which perhaps parallels the secret instructions of Christ to members of his council. The book is informative, inspiring, and splendidly written. Paperback.

MENTAL POISONING
THOUGHTS THAT ENSLAVE MINDS

by H. Spencer Lewis, F.R.C., Ph. D.

Must humanity remain at the mercy of evil influences created in the minds of the vicious? Do poisoned thoughts find innocent victims? Use the knowledge this book fearlessly presents as an antidote for such superstitions and their influences.

There is no need to remain helpless even though evil thoughts of envy, hate, and jealousy are aimed to destroy your self-confidence and peace of mind.

GLANDS—THE MIRROR OF SELF

by Onslow H. Wilson, Ph.D., F.R.C.

You need not continue to be bound by those glandular characteristics of your life which do not please you. These influences, through the findings of science and the mystical principles of nature, may be adjusted. The first essential is that of the old adage, "Know Yourself." Have revealed to you the facts about the endocrine glands—know where they are located in your body and what mental and physical functions they control. The control of the glands can mean the control of your life. These facts, scientifically correct, with their mystical interpretation, are presented in simple, nontechnical language, which everyone can enjoy and profit by reading.

THE SANCTUARY OF SELF

by Ralph M. Lewis, F.R.C.

Are you living your life to your best advantage? Are you beset by a *conflict of desires*? Do you know that there are various *loves* and that some of them are dangerous drives?

Learn which of your feelings to discard as enslaving influences and which to retain as worthy incentives.

The author, Imperator of the Rosicrucian Order, brings to you from his years of experience, the practical aspects of mysticism.

SEPHER YEZIRAH—A BOOK ON CREATION
or The Jewish Metaphysics of Remote Antiquity

by Dr. Isidor Kalisch, Translator

The ancient basis for Kabalistic thought is revealed in this outstanding metaphysical essay concerning all creation. It explains the secret name of Jehovah.

Containing both the Hebrew and English texts, its 61 pages have been photolithographed from the 1877 edition. As an added convenience to students of Kabala, it contains a glossary of the original Hebraic words and terms.

SON OF THE SUN

by Savitri Devi

The amazing story of Akhnaton (Amenhotep IV), Pharaoh of Egypt, 1360 B.C. This is not just the fascinating story of one life—it is far more. It raises the curtain on man's emerging from superstition and idolatry. Against the tremendous opposition of a fanatical priesthood, Akhnaton brought about the world's first spiritual revolution. He was the first one to declare that there was a "sole God." In the words of Sir Flinders Petrie *(History of Egypt)*: "Were it invented to satisfy our modern scientific conceptions, his religio-philosophy could not be logically improved upon at the present day."

This book contains over three hundred pages. It is handsomely printed, well bound, and stamped in gold.

THE CONSCIOUS INTERLUDE

by Ralph M. Lewis, F.R.C.

With clarity of expression and insightful penetration of thought, this original philosopher leads us to contemplate such subjects as: the Fourth Dimension, the Mysteries of Time and Space; the Illusions of Law and Order; and many others of similar import.

As you follow the author through the pages into broad universal concepts, your mind too will feel its release into an expanding consciousness.

ESSAYS OF A MODERN MYSTIC

by H. Spencer Lewis, F.R.C., Ph. D.

These private writings disclose the personal confidence and enlightenment that are born of *inner experience.* As a true mystic-philosopher, Dr. Lewis shares with his readers the results of contact with the Cosmic Intelligence residing within.

COSMIC MISSION FULFILLED

by Ralph M. Lewis, F.R.C.

This illustrated biography of Harvey Spencer Lewis, Imperator of the Ancient Mystical Order Rosae Crucis, was written in response to the requests of thousands of members who sought the key to this mystic-philosopher's life mission of rekindling the ancient flame of *Wisdom* in the Western world. We view his triumphs and tribulations from the viewpoint of those who knew him best.

Recognize, like him, that the present is our *moment in Eternity*; in it we fulfill our mission.

WHISPERINGS OF SELF

by Validivar

Wisdom, wit, and insight combine in these brief aphorisms that derive from the interpretation of Cosmic impulses received by Validivar, whose true name is Ralph M. Lewis, Imperator of the Rosicrucian Order.

These viewpoints of all areas of human experience make an attractive gift as well as a treasured possession of your own.

HERBALISM THROUGH THE AGES

by Ralph Whiteside Kerr, F.R.C.

The seemingly magical power of herbs endowed them with a divine essence to the mind of early man. Not only did they provide some of his earliest foods and become medicines for his illnesses but they also symbolized certain of his emotions and psychic feelings. This book presents the romantic history of herbs and their use even today.

ETERNAL FRUITS OF KNOWLEDGE

by Cecil A. Poole, F.R.C.

A stimulating presentation of philosophical insights that will provoke you into considering new aspects of such questions as: the purpose of human existence, the value of mysticism, and the true nature of good and evil. Paperback.

CARES THAT INFEST...

by Cecil A. Poole, F.R.C.

With a penetrating clarity, Cecil Poole presents us with the key to understanding our problems so that we may open wide the door and dismiss care from our lives. The author guides us on a search for *true value* so that, in the poet's words, "the night will be filled with music," as the *cares* "silently steal away."

MENTAL ALCHEMY

by Ralph M. Lewis, F.R.C.

We can transmute our problems to workable solutions through *mental alchemy*. While this process is neither easy nor instantaneously effective, eventually the serious person will be rewarded. Certain aspects of our lives *can* be altered to make them more compatible with our goals.

Use this book to alter the direction of your life through proper thought and an understanding of practical mystical philosophy.

MESSAGES FROM THE CELESTIAL SANCTUM

by Raymond Bernard, F.R.C.

The real *unity* is Cosmic Unity. You are never separated from the Cosmic, no matter where you live or how different your lifestyle may be. Each person is like a channel through which cosmically inspired intuitive impressions and guidance can flow. This book explains how you can harmonize yourself with the *Celestial Sanctum,* an all-encompassing phenomenon that reveals rational, sensible, and practical messages of a cosmic nature. Allow this book to show you how your mind can become a window through which you can observe creation—and learn from it in a *personal way*.

IN SEARCH OF REALITY

by Cecil A. Poole, F.R.C.

This book unites metaphysics with mysticism. Man is not just an isolated entity on Earth. He is also of a great world—the Cosmos. The forces that create galaxies and island universes also flow through man's being. The human body and its vital phenomenon—Life—are of the same spectrum of energy of which all creation consists. The universe is you because you are one of its myriad forms of existence. Stripping away the mystery of this Cosmic relationship increases the personal reality of the Self. Paperback.

THROUGH THE MIND'S EYE

by Ralph M. Lewis, F.R.C.

Truth is what is real to us. Knowledge, experience, is the material of which truth consists. But what is the *real, the true,* of what we know? With expanding consciousness and knowledge, truth changes. Truth therefore is ever in the *balance*—never the same. But in turning to important challenging subjects, the *Mind's Eye* can extract that which is the true and the real, for the *now.* The book, *Through The Mind's Eye,* calls to attention important topics for judgment by your mind's eye.

MYSTICISM—THE ULTIMATE EXPERIENCE

by Cecil A. Poole, F.R.C.

An experience is more than just a sensation, a feeling. It is an *awareness,* or perception, with *meaning.* Our experiences are infinite in number, yet they are limited to certain types. Some are related to our objective senses; others, to dreams and inspirational ideas. But there is *one* that transcends them all—the *mystical experience.* It serves every category of our being: it stimulates, it enlightens, it strengthens; it is the *Ultimate Experience.*

And this book, *Mysticism—The Ultimate Experience,* defines it in simple and inspiring terms.

THE CONSCIENCE OF SCIENCE
and Other Essays

by Walter J. Albersheim, Sc.D., F.R.C.

A remarkable collection of fifty-four essays by one of the most forthright writers in the field of science and mysticism. His frank and outspoken manner will challenge readers to look again to their own inner light, as it were, to cope with the ponderous advances in modern technology.

THE UNIVERSE OF NUMBERS

From antiquity, the strangest of systems attempting to reveal the universe has been that of numbers. This book goes back to the mystical meaning and inherent virtue of numbers. It discusses the kabalistic writings contained in the *Sepher Yezirah,* and correlates the teachings of Pythagoras, Plato, Hermes Trismegistus, Philo, Plotinus, Boehme, Bacon, Fludd, and others who have explored this fascinating subject. *(Formerly published as "Number Systems and Correspondences.") Paperback.*

GREAT WOMEN INITIATES or The Feminine Mystic

by Hélène Bernard, F.R.C.

Throughout history, there have been women of exceptional courage and inspiration. Some, such as Joan of Arc, are well known; others have remained in relative obscurity—until now. In this book, Hélène Bernard examines from a Rosicrucian viewpoint the lives of thirteen great women mystics. Her research and insight have unveiled these unsung heroines who, even in the face of great adversity, have staunchly defended freedom of thought and the light of mysticism. Paperback.

INCREASE YOUR POWER OF CREATIVE THINKING IN EIGHT DAYS

by Ron Dalrymple, Ph.D., F.R.C.

Most people know they have the power within themselves to speak out and to do wonderful things, but they are constantly frustrated by not knowing how to get those great ideas out in the open. This new workbook contains simple but effective techniques designed to help you think more creatively. Its step-by-step program of exercises helps stimulate the flow of creative ideas by helping you develop a creative attitude and learn to think in creative patterns. Paperbound.

THE IMMORTALIZED WORDS OF THE PAST

Take a fascinating journey of the mind and spirit as you read these inspired writings. Fifty-eight of the world's most courageous thinkers bring you the benefit of their knowledge and experiences. Each excerpt is accompanied by a biographical sketch of its author. From Ptah-hotep to Albert Einstein, discover the wisdom of those who pioneered the highest avenues of human expression.

A SECRET MEETING IN ROME

by Raymond Bernard, F.R.C.

Experience a mystical quest for knowledge through this allegorical story of initiation. Narrated in the first person, this book symbolically explains the modern mission of the Order of the Temple and its connection with Atlantis, Pharaoh Akhnaton, and the Rose-Croix. You will also learn about the esoteric relationship between Christianity and Islam and the search for the Holy Grail. The author sheds light upon these mysteries and more as his dramatic story unfolds.

THE MYSTIC PATH

by Raymund Andrea, F.R.C.

Here is an informative and inspirational work that will guide you across the threshold of mystical initiation. The author provides insights into the states of consciousness and experiences you may have as you travel the Mystic Path. It is filled with the fire and pathos of the initiate's quest. His spiritual, mental, and physical crises are fully described and pondered. Andrea's deep understanding of the essence of Western mystical and transcendental thought makes this a book you will treasure and refer to often as you advance in your mystical studies. Among the many topics addressed are: Meditation, Contemplation, Awakening Consciousness, the Dark Night of the Soul, Mystical Participation, and Mystical Union.

COMPASS OF THE WISE

by Ketmia Vere
translated by Léone Muller

The extraordinary works of sixteenth-century Rosicrucian, Michael Maier, and many other mystic alchemist-philosophers contributed to this 18th-century Rosicrucian book. This rare manuscript, first published in Germany in 1779 and newly translated into English from that original German text, will be of import to Rosicrucians, Freemasons, and others interested in alchemy. Included is a brief history of the Rosicrucian Order, while the majority of the book is devoted to alchemy—the theory of its processes and points about its practice. This is an invaluable work for anyone deeply interested in understanding the processes of physical and spiritual alchemy as expressed in the Rosicrucian tenet: "As above, so below."